Svelte 3
Up and Running

A fast-paced introductory guide to building
high-performance web applications with SvelteJS

Alessandro Segala

BIRMINGHAM—MUMBAI

Svelte 3 Up and Running

Copyright © 2020 Packt Publishing

Commissioning Editor: Pavan Ramchandani

Acquisition Editor: Rohit Rajkumar

Senior Editor: Richard Brookes-Bland

Content Development Editor: Keagan Carneiro

Technical Editor: Deepesh Patel

Copy Editor: Safis Editing

Project Coordinator: Kinjal Bari

Proofreader: Safis Editing

Indexer: Tejal Daruwale Soni

Production Designer: Jyoti Chauhan

First published: August 2020

Production reference: 2241120

Published by Packt Publishing Ltd.

Livery Place

35 Livery Street

Birmingham

B3 2PB, UK.

ISBN 978-1-83921-362-5

www.packt.com

To those who made this book possible:

Maria Grazia, Francesco, and Clare, for their continued support.

All the people who contribute to open source projects and build incredible things like Svelte.

– Alessandro Segala

`Packt.com`

Subscribe to our online digital library for full access to over 7,000 books and videos, as well as industry leading tools to help you plan your personal development and advance your career. For more information, please visit our website.

Why subscribe?

- Spend less time learning and more time coding with practical eBooks and Videos from over 4,000 industry professionals

- Improve your learning with Skill Plans built especially for you

- Get a free eBook or video every month

- Fully searchable for easy access to vital information

- Copy and paste, print, and bookmark content

Did you know that Packt offers eBook versions of every book published, with PDF and ePub files available? You can upgrade to the eBook version at `packt.com` and as a print book customer, you are entitled to a discount on the eBook copy. Get in touch with us at `customercare@packtpub.com` for more details.

At www.`packt.com`, you can also read a collection of free technical articles, sign up for a range of free newsletters, and receive exclusive discounts and offers on Packt books and eBooks.

Contributors

About the author

Alessandro Segala is a Product Manager at Microsoft working on developer tools. He has over a decade of experience building full-stack web applications, having worked as a professional developer as well as contributing to multiple open source projects. Alessandro is the maintainer of svelte-spa-router, one of the most popular client-side routers for Svelte 3.

About the reviewer

Russell Jones is an enthusiastic software engineer, who spends most of his time programming in Javascript. He is a Founder of Functional, Inc and has been building web applications since the 1990s. Russ lives in Portland, Maine, with his two children, Ailee and Coda, who are his greatest engineering achievements. Pre-COVID, Russ spent most of his time in coffeeshops smiling at the locals and making keyboard noises in the corner. These days, Russ simply spends too much time playing video games.

Packt is searching for authors like you

If you're interested in becoming an author for Packt, please visit `authors.packtpub.com` and apply today. We have worked with thousands of developers and tech professionals, just like you, to help them share their insight with the global tech community. You can make a general application, apply for a specific hot topic that we are recruiting an author for, or submit your own idea.

Table of Contents

3

Building Reactive Svelte Components

4

Putting Your App Together

5

Single-Page Applications with Svelte

Preface

Svelte is a modern JavaScript framework used to build static web apps that are fast, lean, and are fun for developers to use. This book is a concise and practical introduction for those who are new to the Svelte framework. It will get you up to speed with building apps quickly, and teach you how to use Svelte 3 to build apps that offer a great user experience (UX).

The book starts with an introduction to Svelte 3, before showing you how to set up your first complete application with the framework. Filled with code samples, each chapter will show you how to write components using the Svelte template syntax and the **Application Programming Interfaces (APIs)** of the Svelte framework. As you advance, you'll go from scaffolding your project and tool setup, all the way to production with DevOps principles such as automated testing, **Continuous Integration**, and **Continuous Delivery (CI/CD)**. Finally, you'll deploy your application in the cloud with object storage services and a **Content Delivery Network (CDN)** for best-in-class performance for your users.

By the end of this book, you'll learn how to build and deploy apps using Svelte 3 to solve real-world problems and deliver impressive results.

Who this book is for

This book is for front-end or full-stack developers who are looking to build modern web apps with Svelte 3. This book assumes a solid understanding of JavaScript and core HTML5 technologies. Basic understanding of modern front-end frameworks (for example Angular, React, Vue) will be beneficial, but not necessary.

What this book covers

Chapter 1, *Meet Svelte,* After covering what modern web development looks like, with concepts such as JAMstack and **Single-Page Apps (SPAs)**, we introduce the Svelte 3 framework and explain how it differs from other popular frameworks like Angular and React: how it can lead to smaller bundles and faster apps. We will also introduce the goal of the book, which is to create a sample app for creating your own journal.

Chapter 2, Scaffolding Your Svelte Project, Before we dive into building the app, we will go through the steps to install all the required tools, such as Node.js and Visual Studio Code (optional). We will then scaffold the project, create the folder structure, and setup Webpack. Lastly, we will create a "hello world" app with Svelte 3 and run it locally with a development server, to verify that our setup is complete and correct.

Chapter 3, Building Reactive Svelte Components, In this chapter, we'll start building the Svelte components that our journaling app uses. In the process, we will learn about the syntax used by Svelte templates, and concepts such as binding and events.

Chapter 4, Putting Your App Together, As we create the last components that our app depends on, we will learn about Svelte stores and transitions. We will then look into other features of the Svelte 3 language and runtime, including a more advanced one.

Chapter 5, Single-Page Applications with Svelte, In the first part of this chapter, we'll focus on building **Single-Page Apps** (**SPAs**) with Svelte 3, including implementing client-side routing (and we'll look at the two main options for doing that, and when to use which). In the second part, we'll briefly look into creating automated tests for our apps using **Nightwatch.js**. Lastly, we'll add **ESLint** to the project and configure it to enforce styling conventions.

Chapter 6, Going to Production, We'll end our journey by bringing the app to production. We'll cover how to serve the app from object storage (Azure Blob Storage) with a CDN in front. We'll also look at pushing our code to GitHub and adding Continuous Integration and Continuous Delivery using GitHub Actions, to automatically build, test, and deploy our app.

Chapter 7, Looking Forward, This chapter looks at what's next for readers, and how they can progress in their knowledge of developing JAMstack apps with Svelte 3.

To get the most out of this book

To build the app described in this book using Svelte 3, you will need a PC or laptop running Windows (7 or higher), a recent version of macOS, or Linux (any commonly used distribution).

In *Chapter 2, Scaffolding Your Svelte Project* as we set up our tooling, we'll guide you through the installation of Node.js and optional tools such as Visual Studio Code.

While Svelte 3 officially requires Node.js 8 or higher, my recommendation is to use the latest **Long-Term Support** (**LTS**) version; at the time of writing, this is version 12.

If you're on Windows, whenever possible we recommend using Windows 10 and leverage the Windows Subsystem for Linux; we'll cover this in the book as well.

If you are using the digital version of this book, we advise you to type the code yourself or access the code via the GitHub repository (link available in the next section). Doing so will help you avoid any potential errors related to the copying and pasting of code.

Download the example code files

You can download the example code files for this book from your account at www.packt.com. If you purchased this book elsewhere, you can visit www.packtpub.com/support and register to have the files emailed directly to you.

You can download the code files by following these steps:

1. Log in or register at www.packt.com.
2. Select the **Support** tab.
3. Click on **Code Downloads**.
4. Enter the name of the book in the **Search** box and follow the onscreen instructions.

Once the file is downloaded, please make sure that you unzip or extract the folder using the latest version of:

- WinRAR/7-Zip for Windows
- Zipeg/iZip/UnRarX for Mac
- 7-Zip/PeaZip for Linux

The code bundle for the book is also hosted on GitHub at https://github.com/PacktPublishing/Svelte-3-Up-and-Running – shortened to https://bit.ly/sveltebook In case there's an update to the code, it will be updated on the existing GitHub repository.

We also have other code bundles from our rich catalog of books and videos available at https://github.com/PacktPublishing/. Check them out!

Conventions used

There are a number of text conventions used throughout this book.

`Code in text`: Indicates code words in text, database table names, folder names, filenames, file extensions, pathnames, dummy URLs, user input, and Twitter handles. Here is an example: "After importing all the requisite modules, we define the `prod` variable, which is `true` if we're building the application for production, as determined from the `NODE_ENV` environmental variable."

A block of code is set as follows:

```
const mode = process.env.NODE_ENV || 'development'
const prod = mode === 'production'
```

When we wish to draw your attention to a particular part of a code block, the relevant lines or items are set in bold:

```
AUTH_CLIENT_ID=00000000-0000-0000-0000-000000000000
API_URL=http://localhost:4343
AUTH_JWKS_URL=http://localhost:4343/jwks
AUTH_URL=http://localhost:4343/authorize?client_
id={clientId}&response_type=id_token&redirect_
uri={appUrl}&scope=openid%20profile&nonce={nonce}&response_
mode=fragment
AUTH_ISSUER=http://svelte-poc-server
KEY_STORAGE_PREFIX=svelte-demo
```

Any command-line input or output is written as follows:

```
$ NODE_VERSION="v12.18.3"
```

Bold: Indicates a new term, an important word, or words that you see onscreen. For example, words in menus or dialog boxes appear in the text like this. Here is an example: "From that page, fetch the **Windows Installer (.msi)** for the LTS version, selecting the correct architecture of your operating system."

Tips or important notes
Appear like this.

Get in touch

Feedback from our readers is always welcome.

General feedback: If you have questions about any aspect of this book, mention the book title in the subject of your message and email us at customercare@packtpub.com.

Errata: Although we have taken every care to ensure the accuracy of our content, mistakes do happen. If you have found a mistake in this book, we would be grateful if you would report this to us. Please visit www.packtpub.com/support/errata, selecting your book, clicking on the Errata Submission Form link, and entering the details.

Piracy: If you come across any illegal copies of our works in any form on the Internet, we would be grateful if you would provide us with the location address or website name. Please contact us at copyright@packt.com with a link to the material.

If you are interested in becoming an author: If there is a topic that you have expertise in and you are interested in either writing or contributing to a book, please visit authors.packtpub.com.

Reviews

Please leave a review. Once you have read and used this book, why not leave a review on the site that you purchased it from? Potential readers can then see and use your unbiased opinion to make purchase decisions, we at Packt can understand what you think about our products, and our authors can see your feedback on their book. Thank you!

For more information about Packt, please visit packt.com.

1
Meet Svelte

When developers think of JavaScript frameworks, the options abound. Even within the narrow scope of front-end development that we're covering in this book, most developers will be familiar with (or have at least heard of) tools such as Angular, React, and Vue.

However, you picked up this book because you've heard of **Svelte**, and you've heard of it's growing in popularity. You've probably also heard that Svelte is different, and that there's something *magical* about it. I'm using the word *magical* on purpose: in fact, the original tagline for the Svelte project was "the magical disappearing UI framework."

This book is about the Svelte 3 front-end JavaScript framework, the last version as of the time of writing, which was released in April 2019.

You can use Svelte to build single, reusable components for projects of any kind, including larger applications written with Angular, React, Vue, or any other frameworks. Or, you can build entire web applications with it – just like we'll be doing in this book.

Among all the various frameworks, in most comparisons, Svelte 3 stands out for its ability to produce smaller code bundles that run faster in the browser, compared to Angular or React. This is a big part of what makes Svelte *magical*. But, perhaps even more importantly, developers largely enjoy working with Svelte 3; building components and applications with it feels to many very similar to using "vanilla JavaScript."

Throughout this book, I'll help you build your first, fully functional project with Svelte 3: a journaling app. We'll go from bootstrapping the development environment all the way to production with an automated continuous integration and delivery pipeline.

While building this sample app, we'll cover the majority of the features of Svelte 3. By the end, you should have a strong foundation to go and create your own apps with Svelte.

In this chapter, we'll cover the following topics:

- Modern web app development
- Why you should use Svelte
- Details on the app we'll be building

Modern web app development

Before we dig deep into Svelte, or any other JavaScript framework for front-end development, it helps to take a quick trip down memory lane and look at how we arrived at modern web app development, and the role that frameworks play.

How the web became static...again

When I built my first website, in 1999, it was just like what you would expect: rich in flashy GIFs, scrolling texts, and eye-hurtlingly bright colors. It was also served by static hosts that were essentially the Italian equivalent of GeoCities.

I had built that website using the most advanced tools available for webmasters at the time: **WYSIWYG** editors (which stands for **What You See Is What You Get**): for me, that meant Microsoft FrontPage.

FrontPage worked just like Wix.com and other similar services of present day, where you build your website visually and the code is generated for you. It differed in that it was a native desktop app and, in the middle of the first browser war, it outputted code that ignored web standards and was optimized for Internet Explorer. Of course, all those websites were fully static. (In case you're wondering, I eventually "graduated" to Macromedia Dreamweaver, which was more standards-compliant; Macromedia was later acquired by Adobe, and Adobe Dreamweaver is still maintained today.)

Ten years later, in 2009, I was still building websites, but then I was making them *dynamic*, with server-side generation in PHP. Those were the days of "rich internet apps," where users could interact with a website, and "web 2.0," centered around collaboration and user-generated content.

Generating web pages on a server was not a novel concept per se: Java EE was released in 1999, Microsoft created ASP in 1996, and the original PHP came out in 1994! However, it was only in the second half of the '00s that those technologies became more accessible to small teams and individual developers, first of all thanks to new, simpler frameworks: for example, Django and Rails were both released in 2005. Additionally, in those years, we started seeing increasingly cheaper options for hosting web applications that require server-side rendering, especially with PHP.

Around the same time, interactivity started to appear in frontends too.

Developers had been building apps that run within browsers for a long time, leveraging plugins such as Java Applets, ActiveX objects, or Macromedia Flash (later, Adobe). Those were plagued with issues ranging from poor performance to serious security flaws, and they always ran almost completely isolated from the rest of the web page and the DOM. JavaScript was mostly considered a tool for designers to add small bits of interactivity to web pages and other gimmicks – just like its creators had originally meant it for.

Then, Google released Gmail in 2004 and the next year, they built Google Maps, and those were among the first large-scale applications to run primarily within the browser itself (any browser) without plugins. They popularized a new way of building web apps, where data is loaded dynamically in response to user action: this technology was then called **Asynchronous JavaScript and XML (AJAX)**. The foundation for AJAX was the `XMLHttpRequest` API, which existed in Internet Explorer as an ActiveX object and was later implemented by other browsers in a compatible way too.

Another relevant change for JavaScript developers at the time was the creation of libraries such as Prototype in 2005 and jQuery the next year. At a time when browsers were still running JavaScript based on the ECMAScript 3 standard, these libraries provided great help by offering convenient features to web developers, such as easier ways to access and manipulate the DOM. They also helped by abstracting away the complexity of targeting all web browsers, including the then-dominant Internet Explorer 6 and its poor support for web standards.

With the help of AJAX and libraries such as jQuery, it became clear to developers that apps could run within a browser itself, and the amount and sophistication of web apps started growing.

Fast forward to 2020, and browser-based web apps are the new norm, to the point where we can often replace native, desktop apps with web ones running inside the client: from spreadsheets, to photo editing tools, to complex 3D video games.

The last decade brought along a set of innovations that helped make front-end development simpler, more accessible, and more powerful.

This included innovations in browsers, with faster JavaScript engines, and in the standards themselves. New HTML5 and CSS3 specifications added a bunch of new features, including the `<video>` tag that killed Adobe Flash for good, the last standing of the plugins. As for JavaScript, ECMAScript 2015 (ES2015, often called ES6) was finally released in 2015, after a much-troubled process (which is an interesting story in itself). Innovation is continuing, with technologies such as WebAssembly posed to completely rethink the way we build apps too.

As for the developer experience, a new class of front-end frameworks started to appear with AngularJS 1.0 being released in 2010, whose goal was to make it easier to write large-scale apps in JavaScript, avoiding the creation of unmaintainable "spaghetti code." We'll look a bit deeper at these frameworks and how they work, and how they compare to Svelte, later in this chapter.

Eventually, in what could be called a clear example of Nietzschean eternal recurrence applied to software development, we're back to building web apps that are completely static, and they have never been more powerful. (We're also using a very large number of GIFs once again, but at least we're doing it ironically this time!)

Building apps with the JAMstack

Along with advancements in the web technology and browsers, developers have been able to leverage new paradigms.

One of the most relevant to us in this book is the so-called **JAMstack**, an acronym of the following:

- JavaScript
- (Reusable) APIs
- (Pre-rendered) Markup

Inside the JAMstack

The JavaScript part shouldn't come too much as a surprise: in the JAMstack, apps are written in JavaScript and run within a web browser. You can interpret this more broadly to refer to all apps that run within a JavaScript VM in a browser, to extend the definition to also include apps that use WebAssembly.

More interesting is the APIs part. One of the most sought-after features of JAMstack apps is how they deliver a great end user experience by allowing interactivity. This is possible thanks to apps interacting with other back-end services through a set of predefined APIs:

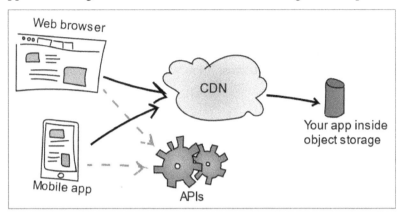

Figure 1.1 - A conceptual diagram for a JAMstack application

The most commonly used APIs are RESTful endpoints that are accessed via HTTP(S), just like we'll be using in this book's sample app. More recently, developers have started to leverage GraphQL too as an alternative format, which is especially optimized for data that can be represented by graphs (GraphQL was invented by Facebook, as a matter of fact). Alternative transport protocols are possible too in certain scenarios, such as using WebSocket for two-way data streaming, or protocol buffers and gRPC for apps that need to transfer large amounts of structured data (although gRPC doesn't yet work natively within web browsers and requires a proxy).

Lastly, pre-rendered markup refers to how JAMstack apps are made of bundles of HTML, JavaScript, and CSS files, as well as assets (images, fonts, and so on) that are rendered at "build-time." Using tools such as Webpack, various source files are packaged together and minified. Text content can also be pre-rendered from Markdown documents, thanks to static site generators such as Gatsby, Hugo, and Jekyll.

JAMstack apps are "compiled" and packaged inside the developers' machines or, more commonly, in a continuous integration server. The final bundle, which contains only HTML, JavaScript, and CSS files, in addition to any static asset, is self-contained, and is often deployed to object storage services in the cloud, such as Azure Blob Storage and AWS S3. Because the applications' files are completely static, developers usually place a **Content Delivery Network** (**CDN**) in front of the object storage service: this caches the files in edge nodes around the world, delivering faster speeds and lower latencies to all visitors worldwide.

Benefits of JAMstack applications

JAMstack apps are enjoying well-deserved popularity, both with developers and end users.

For teams building and operating apps, the benefits of the JAMstack mainly fall into two categories: simpler operations and better developer experience.

The latter group enjoys a great user experience, being able to use apps that are interactive and yet run within a web browser. But, most importantly, JAMstack apps *feel* fast.

For operators

On one hand, operations are simpler and cheaper than for traditional web applications:

- Because JAMstack apps are just a bundle of static files, they can be deployed to object storage services on the cloud, which offer high availability and reliability, and are shockingly inexpensive: their billing model normally charges a few cents per GB stored per month. This is in addition only to the egress bandwidth cost, which you would be charged for regardless of the way the application is built.

- When deploying the application to object storage services, there's no need to maintain a complex infrastructure: teams don't need to deal with things such as containers and container orchestrators such as Kubernetes. There are also no servers to maintain, nor operating systems and application frameworks to patch.

- Because the apps maintain no state, replication, including geo-redundancy, involves just copying the file bundles to the separate locations and using a bare-bones load balancer.

- JAMstack apps can be deployed atomically by copying the new bundle of "compiled" files in a staging directory, and pointing the application to the new location, with minimum downtime.

- Lastly, as mentioned previously, apps can be cached by CDNs completely and effectively, offering faster speeds and lower latencies to your end users.

As a nice side benefit, serving your application's front-end and all static assets (scripts, images, and so on) from an object storage service and a CDN reduces the load on your infrastructure, which receives requests only for the data.

For developers

On the other hand, JAMstack apps come with a delightful **Developer Experience** (**DX**) too.

You might have noticed a similarity between the way JAMstack apps are designed and mobile or desktop ones. In both cases, the application is self-contained and runs entirely within the client, and it communicates with back-end services over the network:

- The first consequence of this is that there's a clear separation of roles and responsibilities. As long as they agree on a common set of APIs, front-end and back-end teams can work autonomously from each other. Both teams also have more freedom in their choices around technologies and stacks, not having to depend on both parties to agree on monolithic frameworks and their templating systems.

- The decoupling of the front-end from the back-end creates the by-product of reusable APIs that can be leveraged by other teams for their purposes, independently. For example, your organization might decide to build a native mobile or desktop app using the same back-end services, while another team could integrate the same APIs in a completely different product, and business users might even build custom interfaces, such as reporting using "no-code" solutions interacting with the same back-end.

- Even the life cycle of the various tiers of the application is decoupled. The front-end team can change their code and redeploy the application without impacting the back-end systems.

- JAMstack apps are normally "compiled" really fast, so developers can see their code's behavior in real time, often with **Hot Module Reload** (**HMR**) support too. The tools required to build and deploy applications are fairly standardized by now, so there are pre-made templates for most continuous integration and continuous delivery platforms.

- Lastly, because the front-end is fully independent from the back-end, frontend developers have the ability to experiment freely. Developers can A/B test various versions of their front-end code independently and safely. In many cases, notwithstanding policies saying otherwise, they might also able to point development front-end apps to production back-end services, with no risk to the availability of the live application.

For end users

As an end user of web applications yourself, this paragraph might be familiar to you.

JAMstack apps feel, and often are, fast. When users perceive your app to be fast, their satisfaction increases, and they are easier to retain.

This is made possible by having a separation between the application code and data, by doing most requests asynchronously and by caching data extensively:

- When requesting a JAMstack app, clients first load the app's *shell*, which consists of the app's code without any data. Because the app's bundle is fully static, this request can easily be served by CDNs: these have plenty of bandwidth and edge nodes in hundreds of places around the world. Users can fetch the app's shell faster, and the reduced distance to the CDN's edge node leads to less latency, which means that they start receiving data sooner. Additionally, browsers can store the app's shell in their own cache, so returning visitors might be able to avoid requesting it completely.

  Even though your app's shell might be many kilobytes in size, thanks to the improved performance of the CDN and the extensive caching, the experience for end users is a positive one.

- Asynchronous loading of the data improves the perceived speed of your application. With JAMstack apps, browsers render the shell first, and then request the data asynchronously. Even as they wait for the data to load, users can see your app's interface and interact with it, making your app *feel* faster overall. This is in contrast with traditional apps, where browsers need to wait for the full HTML, both structure and content, before they can render the page and accept interactions.

- Because your server doesn't have to render full-blown HTML pages, page generation time is faster. But even more importantly, the amount of data transferred to users is smaller.

- Even more, by loading data asynchronously, apps can prioritize the content they're loading. For example, you could request and render the content for your app's main view before populating the sidebar.

Compared with desktop and mobile applications, JAMstack apps are delivered via a web browser using the **Software-as-a-Service** (**SaaS**) model, with constant updates, fixes, and new features, to the delight of your users.

Third-party APIs

I mentioned that the API part of the JAMstack is the most interesting one: in my opinion, this is where the biggest opportunities are.

In fact, as apps' tiers become more decoupled, you might even be able to build JAMstack apps that do not integrate with any back-end services that you (or your organization) manage.

Identity services

Perhaps the best example of this is authentication, where your app can integrate with external identity providers. There are multiple reasons why I consider this a good idea rather than rolling out your own identity system, starting from the obvious: that you have less code to manage – especially code that is usually outside your core business logic – to the fact that larger, more robust identity services can offer significant security benefits.

If you're building an enterprise application, it's likely that your organization's identities are already synchronized with Azure Active Directory (Azure AD) (for organizations that use Microsoft 365) or G Suite, both of which can be used by third-party applications too.

For consumer apps, using a social identity is often a convenient choice: Apple, Facebook, GitHub, and so on.

There are also other services, such as Auth0, Okta, and Azure AD B2C, that offer flexible, powerful options, including support for built-in accounts (users can sign up and create a new profile for your app) as well as integration with external services.

SaaS services

Aside from authentication, you can integrate with a very large number of APIs offered by SaaS providers.

For enterprise applications, you can get a vast amount of data by integrating with your organization's productivity suite, such as Microsoft 365 or G Suite. For example, you can access your users' calendars, store data inside their cloud storage space (OneDrive for Business or Google Drive), send and receive emails, create and share presentations and spreadsheets, make phone calls, and much more.

Consumer apps can leverage services such as Dropbox, Google Drive, or Microsoft OneDrive to store arbitrary, persistent data.

Besides that, you can find APIs online for literally anything, from showing maps to providing directions, from searching for images to collecting payments, and from tracking parcels to getting the status of a flight.

APIs for developers and integrations

In this far-from-exhaustive list of APIs, there are two more sub-categories of services that you could leverage in your JAMstack apps.

The first one is about APIs specifically meant for developers looking at integrating features in their applications. Examples of real API services include the following:

- Creating a thumbnail for an image

- Video encoding

- Converting files into other formats, such as creating PDFs from a Word document, or converting PNG images into JPG

- Detecting fraudulent activities

- Stopping bots – for example, with CAPTCHAs

- Triggering a software release with a continuous delivery pipeline

- …and many more!

The other group is about using "low-code" or "no-code" platforms to perform certain automated tasks for your application. Services in this category include Microsoft Power Automate, Azure Logic Apps, and Zapier.

These even allow connecting to resources (such as databases, ERP systems, and so on) that cannot otherwise be safely accessed by a front-end only app or performing actions automatically in response to events; for example, sending an email notification every time you receive a new tweet or recording an entry in a spreadsheet when someone presses a button in your web app.

In the consumer space, IFTTT offers a large amount of integrations, including social networks, weather forecasts, smart home appliances (for example, lightbulbs, thermostats, refrigerators, and so on), personal fitness solutions, and other consumer platforms and services.

Why use external APIs?

The most obvious reason why you might want to integrate external APIs is that you don't have to manage them. It becomes someone else's responsibility to ensure that they're available and scale, to apply security fixes, and so on.

Another benefit might be around compliance. For example, using an external payment platform, such as Stripe or Square, frees developers from having to build applications and infrastructures that comply with PCI-DSS. If your app doesn't store any user data, you might also have an easier time adhering to privacy regulations, such as GDPR, because the burden of compliance falls on the API service providers. (But confirm this with your legal team.)

JAMstack versus SPAs and PWAs

At this point, you might be wondering how the JAMstack compares with other trends in web app development, namely **Single-Page Applications** (**SPAs**) and **Progressive Web Apps** (**PWAs**).

SPAs are web applications where all the views are contained in a single HTML page, and routing happens inside the browser. This is in contrast to multi-page applications, where each view has its own HTML page, and clients navigate between views by requesting a different page from the server.

PWAs are web applications that have three features: they use HTTPS, have a manifest file that makes them "installable" in a client, and leverage the service workers APIs in the browser to cache data and provide an offline-enabled experience to your users.

All of these three kinds of apps (single- and multi-page apps and PWAs) can be built with the JAMstack, as long as they fit the definition. That means having the requirements of being built with JavaScript (**J**), interacting with reusable APIs (**A**), and using pre-rendered markup (**M**). In other words, they're JAMstack apps if they can be exported as static files and don't require server-side generation.

> **Important note**
>
> The demo app we're building is a SPA. However, you can use Svelte 3 to build any kind of app, including multi-page apps and PWAs.
>
> Additionally, Svelte 3 can be used to pre-render content in a server-side application, so it could be used to build non-JAMstack apps. We will not cover such an advanced scenario in this book.

Why Svelte?

To understand what's special about Svelte, we need to look at how it differs from other popular libraries and frameworks, such as Angular, React, and Vue.

Svelte versus the other frameworks

Svelte's goal is to help developers write less code, letting them build components using familiar HTML, CSS, and JavaScript. Just like React ones, Svelte apps are truly reactive, so you do not need to manipulate the DOM directly (as you would if you were using jQuery, for example): the view is automatically updated on every change in the state.

However, Svelte's main intuition, and its biggest difference from all the other popular JavaScript frameworks, is that it moves most of the processing to a **compilation stage**. That is, rather than relying on large and complex libraries loaded at run time, Svelte is built around a compiler that processes your application's code before emitting a small, fast, and optimized code bundle.

Another major difference from other popular frameworks is that Svelte requires almost no boilerplate code: Svelte components are written with HTML, CSS, and JavaScript blocks. Scripts feel very close to "vanilla JavaScript," while markup is defined using standard HTML tags. Svelte does not introduce new JavaScript syntax either, unlike React's JSX.

Thanks to being pre-compiled, Svelte apps have a minimal overhead, both in terms of bundled code size and performance. Smaller bundles make your pages load faster, especially for users with slower internet connections. Additionally, by performing the bulk of the work in the compilation stage, Svelte does not need to include techniques such as the virtual DOM used by React and Vue: state changes in Svelte apps are reflected directly in the DOM, without extra overhead.

In a sense, Svelte is a response to the explosion of the size of web pages. According to the HTTP Archive reports (`https://www.httparchive.org/reports/state-of-the-web?start=2015_03_01&end=latest&view=list`), the median size of a web page was 1,280 KB in 2015, and that has grown to 2,080 KB in 2020. Large web pages take longer to load, increasing the likelihood of users leaving your site. But they also disproportionally impact users in rural areas, where internet bandwidth is more limited.

While the requirement for compiling your JavaScript code might sound off-putting, Svelte does that in a very transparent way, by working directly within bundlers such as Webpack and Rollup. In that sense, it's not much different from other tasks that JavaScript developers perform habitually, such as transpiling to an older ECMAScript version with Babel, or converting type-safe TypeScript code into executable JavaScript. The compilation stage for Svelte is really fast, adding an unnoticeable amount of work to a normal bundling pipeline.

The Svelte project

Svelte is an open source project, whose source code is available on GitHub and was released under a permissive MIT license.

> **Relevant links:**
>
> Official website: `https://svelte.dev/`
> Project page on GitHub: `https://github.com/sveltejs/svelte`

Compared with the likes of Angular and React, Svelte is a relatively recent framework. It was originally created in 2016 by Rich Harris (`https://github.com/Rich-Harris`), a software developer and visual journalist.

Svelte was first noticed by the broader JavaScript developer community in the summer of 2018, when Harris presented it in a talk at JSConf EU. This was the first time I'd heard of it myself, with an introduction to version 2.

The release of version 3 in April 2019 marked an important moment for Svelte (this is still the most recent branch as of the time of writing), as it was a complete rewrite that introduced a new, simpler syntax for writing Svelte components. With Svelte's third release also came a real uptick in the adoption of the framework.

Today, Svelte is maintained by an active community that is organized around the project's GitHub repository, and builds new releases with bug fixes and new features regularly.

It is also used by hundreds of organizations around the world, including The New York Times (where Harris is employed), 1Password, Rakuten, Philips, and GoDaddy. A list of users is maintained on the project's website.

As for developers' preferences, the 2019 State of JavaScript report (`https://2019.stateofjs.com/front-end-frameworks/`) portrays Svelte as well-positioned:

- 88% satisfaction among developers; that makes Svelte the second most popular framework behind React (with 89% satisfaction).

- Svelte ranked first among developers' interest, at 67%.

- However, Svelte's relatively young age made 2019 the first time it appeared in the State of JavaScript report, and a quarter of respondents stated they had never heard of it.

- Nevertheless, a fifth of Svelte users reported working in a company with over 1,000 employees (the same as React), indicating how the framework has gained momentum in more mature organizations too.

Building a journaling app

Throughout the following chapters of this book, we'll build a proof-of-concept application using Svelte 3 – a journaling app:

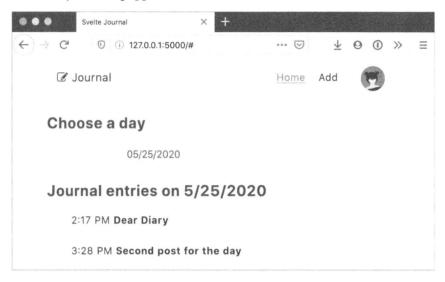

Figure 1.2 – Screenshot of the completed app

App features

This app features three main views:

- The starter view is a list of all the journal entries for a given day; a date picker lets users select the day.

- Users can add new content using a form that lets them type it in freely.

- Content is presented to users when they select a journal entry, rendering the input text as Markdown.

The app requires authentication before users can read or write any journal entries, and it uses OAuth 2.0/OpenID Connect to achieve that.

Data is stored inside a back-end service, which runs separately from our application (remember that we're building a JAMstack app!), and our front-end communicates with the back-end service via RESTful APIs.

While this proof-of-concept app has simplified, limited functionality, it does help us learn all the core capabilities of Svelte 3 (and a few other things too). The concepts we'll be exploring are applicable both to internal, line-of-business applications, as well as external-facing ones.

The back-end service

In order to keep this book focused on front-end development using Svelte 3, we'll be using a pre-built back-end service that provides persistent storage for the data, as well as identity and access control.

To make your life simple and to let you focus on the front-end, I've built a small service that provides the required endpoints. This is written in Go and is available on GitHub at `PacktPublishing/Svelte-3-Up-and-Running` or `https://bit.ly/sveltebook`. You can download the pre-compiled binary for Windows, macOS, and Linux, and launch it on your laptop (usually by double-clicking on the executable) to instantly have the required APIs available for your front-end app to interact with.

Besides being a cheap workaround to the complexity of requiring you to build a back-end, this approach reflects the way modern web applications are built. As the developer working on the front-end portion of your app, you will need to interact with a service using predefined APIs that are maintained by a different team.

You might not have knowledge of how the back-end service works (and, truthfully, you might not want to either), and the service might be written in a completely different stack, just like this back-end app is written in Go. As we saw in the previous sections, in fact, many JAMstack applications interact with services that are maintained by completely different organizations within the company, or even different companies.

Important note

The source code for the back-end service, written in Go, is available on GitHub for you to look at and modify as you please.

The service includes common RESTful endpoints to store, retrieve, and search objects (journal entries).

It also features a mock OAuth 2.0/OpenID Connect implementation to provide identity services. This was built from scratch, but it includes just the bare minimum features to support the needs of the sample front-end application.

While the back-end service is functional, because its purpose is just to aid the development of the front-end application in this book, it is full of sub-optimal practices. In short, **do not use this app or any of its code in production as is**. This is especially key for the access management part, which is likely unsafe for any real-world applications; instead, you should rely on your organization's directory or, if building a consumer-facing app, on trustworthy identity service providers.

Summary

In this first chapter, we learned about the Svelte 3 framework and what makes it *magical* compared to other alternatives for front-end development.

We also did a retrospection on modern web app development, looking at concepts such as JAMstack apps and their benefits to developers and end users.

Lastly, we looked at a description of the application we'll be building throughout this book, which is a journaling app running entirely within the web browser and developed with the Svelte 3 framework.

In the next chapter, we'll start getting our hands dirty and begin building the sample app. We'll begin by setting up all the required tools and scaffolding for our project. At the end, we'll be ready to run a "hello world" app with Svelte 3.

2
Scaffolding Your Svelte Project

After hopefully selling you on the benefits of building your next project as a JAMstack app with Svelte 3, we're now ready to start developing our proof-of-concept application.

However, as with every new technology or framework, we first need to prepare our environment to be ready to work with it. This requires installing the necessary interpreter and runtime dependency, setting up tooling, and scaffolding a new project.

In this chapter, we'll be covering the following topics:

- Setting up your environment: We'll install Node.js, set up Visual Studio Code as our editor (optional, but recommended), and start the backend service for our app.

- Scaffolding a project with Webpack: We'll create our Svelte project and scaffold it using Webpack as a bundler.

- "Hello, Svelte!": Our take on "hello world".

- Debugging Svelte applications in a web browser.

At the end of this chapter, you'll have a scaffolded project for Svelte 3. You can find the end result on GitHub at `https://bit.ly/sveltebook-ch2`.

Setting up your environment

Before we start coding, we need to ensure that we have everything we need installed and ready to use.

The only thing that is absolutely necessary in order to use Svelte is to have Node.js installed. The Svelte compiler itself is written in JavaScript, and it requires Node.js to be executed. Additionally, our development toolchain (the Webpack bundler and all the other build tools) run on Node.js, too.

For our sample application to work, we will also need a back-end service, as mentioned in the previous chapter, which we can run on our development machine.

Installing Node.js

If you don't have it already, you need to install the Node.js framework in your development machine.

While Svelte 3 officially requires Node.js 8 or higher, my recommendation is to use the latest **Long-Term Support (LTS)** version; at the time of writing, this is version 12.

The default Node.js installation comes with NPM too, which is the main package manager for JavaScript applications. Our Svelte project will heavily depend on NPM, too.

Installing on Windows

On Windows, the easiest approach is to install Node.js using the official installer downloaded from `https://nodejs.org/en/download/`.

From that page, fetch the **Windows Installer (.msi)** for the LTS version, selecting the correct architecture of your operating system (64-bit is the most common one at the time of writing). After downloading the installer, launch the application and follow the installation instructions on screen. The official installer includes NPM too. Note that the installer requires Administrator privileges, and you might need to authenticate with the system.

> **Using WSL on Windows 10**
>
> If you're on Windows 10, you have the option of using the **Windows Subsystem for Linux (WSL)** tool, which lets you run a full-fledged Linux environment within your Windows machine, and it is deeply integrated with Visual Studio Code. Many Node.js developers prefer to use WSL on Windows as some packages on NPM are not optimized for Windows.
>
> If you decide to use WSL, you first need to enable it on Windows 10. Microsoft has published the necessary instructions at `https://aka.ms/wsldocs`. After WSL is installed and you have a Linux distribution ready, follow the instructions for Linux in the following section.

Installing on macOS

Just like on Windows, on macOS, you can install Node.js and NPM with the official installer published on `https://nodejs.org/en/download/`.

From that page, fetch the **macOS Installer (.pkg)** for the LTS version, which includes NPM too. After downloading it, launch the installer and follow the steps on screen.

Installing on Linux

On Linux, installing Node.js varies according to your distribution.

The Node.js project publishes pre-compiled binaries for Linux, for both x64 (64-bit Intel-compatible CPU) and ARM (both 32-bit and 64-bit). However, these require manual installation.

While most Linux distributions do ship official packages with Node.js, they're often very outdated, and I do not recommend using them.

Installing using a package manager

NodeSource maintains "semi-official" packages for the most popular Linux distributions. While these are not maintained by the Node.js project or the OpenJS Foundation, a link to these packages is available on the official Node.js website, too, and they're arguably the most popular option for Linux users.

These packages are available in the following forms:

- DEB packages, which is the format supported by Debian, Ubuntu, and other Debian-based distributions, including Linux Mint and Raspberry Pi OS.

- RPM packages, which is the format supported by Red Hat Enterprise Linux, CentOS, Fedora, Amazon Linux, and distributions based on those.

- Snaps for all distributions that support snapd.

You can find the list of packages and installation instructions on GitHub: `https://github.com/nodesource/distributions`

Installing from the official binaries

As mentioned, the Node.js project publishes official binaries for Node.js for all architectures, but they come without an installer.

Nevertheless, using the binary tarball (a `.tar.gz` archive) can give you more control over the version(s) of Node.js that you run, as well as more flexibility.

To install Node.js using the official binary tarball, I recommend uncompressing it into a folder such as `/usr/local/node-$NODE_VERSION-linux-x64` and then creating a symbolic link to that folder from `/usr/local/node`. Not only will this give you the flexibility to update Node.js by downloading a new tarball and changing the symbolic link, but it will also minimize the chances that another package you install from your distribution's repositories brings in a conflicting binary.

To start, set the version of Node.js you want to use in an environmental variable for the next commands. As of the time of writing, the latest LTS version available is 12.18.3, but check on `https://nodejs.org` first:

```
$ NODE_VERSION="v12.18.3"
```

Then, download and uncompress the binary tarball:

```
$ curl -LO http://nodejs.org/dist/$NODE_VERSION/node $NODE_
    VERSION-linux-x64.tar.gz
$ tar xzf node-$NODE_VERSION-linux-x64.tar.gz
$ sudo cp -rp node-$NODE_VERSION-linux-x64 /usr/local/
$ sudo ln -s /usr/local/node-$NODE_VERSION-linux-x64 /usr/
    local/node
```

After this, the Node.js binary is available at `/usr/local/node/bin/node`.

You can add `/usr/local/node/bin` to your `$PATH` to be able to invoke `node` directly. The instructions for doing this depend on the shell you use and on the location of your shell's profile file (for example `.bash_profile` or `.zprofile`). Normally, after identifying the correct file, you'd append the following line to it:

```
export PATH=$PATH:/usr/local/node/bin
```

Installing using NVM

For macOS and Linux (including WSL on Windows 10), an alternative way to install Node.js and NPM is to use NVM, or the Node Version Manager.

NVM is a set of scripts that allow Node.js to be installed and updated, multiple versions maintained, and even different versions of Node.js specified on a per-project basis.

Full reference to NVM and all its options can be found on the project's GitHub page: `https://github.com/nvm-sh/nvm`

Here are the commands to install NVM on a macOS, Linux, and WSL environment, selecting version 12:

```
$ curl -o- https://raw.githubusercontent.com/nvm-sh/nvm/
    v0.35.3/install.sh | bash
$ nvm install 12
```

Verifying that Node.js and NPM are installed

Regardless of your platform and operating system, to test whether Node.js and NPM are installed correctly, you can open a terminal and run the following command:

```
$ node -v
v12.18.3
$ npm -v
6.14.5
```

You should see the versions of Node.js and NPM that you just installed, just like in my example above (your versions might be newer than mine).

Setting up Visual Studio Code

Let me begin this section by saying that this is fully optional, and you're welcome to use your favorite code editor instead.

However, while all editors are good, Visual Studio Code is probably the most convenient option for working with Svelte, thanks to having specific extensions for it that offer syntax highlighting and checking, autocompletion, and so on.

If you're not familiar with Visual Studio Code, it's a free and open source editor from Microsoft that runs on Windows, macOS, and Linux, and it is highly customizable thanks to its vast extension marketplace. Visual Studio Code is the most popular code editor according to the StackOverflow 2019 Developer Survey, and it's especially popular among developers working with JavaScript.

To start with Visual Studio Code, follow these steps:

1. Download the installer for your platform from the official website: `https://code.visualstudio.com`. Then, install it by following the usual process for your operating system or distribution.

2. After installing Visual Studio Code, install the free Svelte extension from the Visual Studio Marketplace: `https://aka.ms/vscode-svelte`. This extension provides syntax highlighting and autocompletion for Svelte components, as well as support for diagnostic messages, and it's strongly recommended for everyone working with Svelte.

3. For debugging within the editor, it's useful to install the **Debugger for Chrome** extension from the Marketplace (similar extensions exist for Firefox and Edge, too): `https://aka.ms/vscode-chrome-debugger`.

4. Lastly, another very useful extension is the NPM one, which is again free from the Visual Studio Marketplace: `https://aka.ms/vscode-npm`. This affords conveniences for using NPM commands directly within the editor, without using a terminal.

> **Note**
>
> If you are working on Windows 10 with Node.js installed on WSL, you might want to configure Visual Studio Code for using WSL, too. The editor should automatically detect the presence of WSL and recommend the extension to you, but full instructions (and gotchas) are available in the documentation: `https://code.visualstudio.com/docs/remote/wsl`

Launching the back-end service

As mentioned in the first chapter, just like in most other real-world situations, our demo application requires a back-end service. This provides REST APIs to access a persistent data store, as well as user authentication.

To keep us focused on working on our Svelte code base, I've built a small application to provide these back-end services. Its source code is available on GitHub (`https://bit.ly/sveltebook`, in the `api-server` folder) and it's distributed as a pre-compiled executable that you can run on your laptop and that works together with the app you're developing.

Running the pre-compiled binary

The simplest way to get the back-end service running is to download the pre-compiled binary from the GitHub project page, and is available for Windows, macOS, and Linux.

Go to `https://bit.ly/sveltebook` and look for **Pre-compiled binaries** links in the `README.md` file, and then download the one for your operating system and architecture. The file is an archive (`tar.gz` or `zip`) that contains the application you need to run.

For most users, launching the application requires just double-clicking on the executable. A terminal window should launch automatically, running a local server.

> **Note for macOS users**
>
> The pre-compiled binary is not signed with an Apple developer certificate, and Gatekeeper will refuse to run it in newer versions of macOS. If this happens to you, you will notice an error saying that the app is coming from an unidentified developer.
>
> To run the application on macOS, you can either (temporarily) disable Gatekeeper and allow unsigned applications (refer to the Apple Support page: `https://apple.co/2E3mVYP`) or run this command: `xattr -rc path/to/application` (where `path/to/application` is the location of the downloaded binary).

The back-end application stores data on your local disk, in a folder called `data` in the same location where you placed the downloaded binary.

To terminate the application, in most cases, you just need to close the terminal window.

Running with Docker

If you have Docker installed in your laptop, you can also launch the back-end service as a container.

This command should work for most users:

```
$ docker run --rm -p 4343:4343 \
    -v ~/data:/data \
    italypaleale/sveltebook
```

This command will launch the Docker container, exposing port 4343 on the local machine for our Svelte application to interact with. It also stores persistent data (for example, your journal entries) in the `data` folder inside your home directory.

To kill the back-end service, press *Ctrl + C* in the terminal window, or close the terminal.

Check whether the application is working

To test whether the back-end application is working correctly, open your web browser and visit the following link:

```
http://localhost:4343
```

If everything is fine, you should see a welcome page:

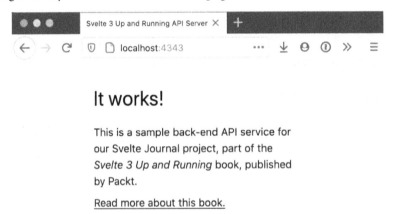

Figure 2.1 – Screenshot of the welcome page in the back-end server application

Scaffolding our project

Now that all the prerequisites are in place, we're ready to start coding:

1. Create an empty folder where you want to put your application's code. In this book, we'll place the code in a folder called `svelte-poc` inside our home directory.

2. Open this folder in Visual Studio Code, or your preferred editor.

3. Because our Svelte application relies on Node.js and NPM, initialize the project by creating a `package.json` file.

    If you have the NPM extension installed in Visual Studio Code, you can launch the command palette with *Ctrl + Shift + P* (or *Command + Shift + P* on a Mac) and then write > `npm run init` and press *Return*.

    Alternatively, using a terminal (tip: you can press *Ctrl +* `, or *Command +* ` on a Mac, to launch an integrated terminal in Visual Studio Code ; where ` is the backtick), run the following command:

    ```
    npm init
    ```

    In both cases, you will be asked to provide some details about your project. At this stage, it's probably safe to accept all defaults by keep pressing the *Return* key until done.

4. Finally, we will need to create two folders inside our project:

    (a) `src`, which contains our application's source

    (b) `public`, which will contain the compiled, bundled code:

Figure 2.2 – What our project looks like right now, with the package.json file and two empty folders

Using a bundler

Writing web applications with Svelte requires the use of a bundler, which merges all JavaScript code into a single file. Not only do they provide convenient features, such as code minification and tree shaking (removing unused code and dependencies), but they also compile Svelte components into executable code. As you will recall from the introduction, in fact, Svelte apps need to be compiled, and that's how Svelte helps to generate small and fast web applications.

The Svelte community is somehow split between two bundlers: Rollup, which also happens to have been originally created by the same Rich Harris who built Svelte, and Webpack. In this book, we're going to make an opinionated choice and use Webpack.

Webpack is arguably the most popular bundler for JavaScript applications. It comes with a host of advanced features (for example, support for code splitting) and a vast number of plugins. Because of its popularity, it's easy to find modules, plugins, documentation, and suchlike, and there's a vibrant community that can provide support. Lastly, using Webpack can allow easier integrations with existing systems. The downside is that Webpack's flexibility makes it a bit more complex, but this book's instructions can help you navigate that complexity.

Scaffolding your project with Webpack

In this section, we'll set up NPM as a bundler.

Installing dependencies from NPM

First, we need to install Webpack and other NPM modules. Use the following command:

```
$ npm install --save-dev \
    webpack@4 \
    webpack-cli@3 \
    webpack-dev-server@3 \
    svelte@3 \
    svelte-loader@2 \
    css-loader@3 \
    style-loader@1 \
    mini-css-extract-plugin@0.9 \
    dotenv-webpack@1 \
    cross-env@7
```

This command installs the preceding modules and adds them to the `package.json` file as `devDependency`, because they are required to build the web application but are not used at runtime. These modules are required only for the bundler, and after you're done scaffolding the project, you won't have to worry about them anymore.

Let's take a look at what they do:

- `webpack`, `webpack-cli`, and `webpack-dev-server` are the three modules required to run Webpack and the built-in web server for development.
- `svelte` and `svelte-loader` are the Svelte framework and compiler, and the loader that enables Webpack to interpret Svelte files.
- `css-loader`, `style-loader`, and `mini-css-extract-plugin` are required to work with CSS styles.
- `dotenv-webpack` is a plugin that allows us to define environmental variables with a `dotenv` file, which we'll use for the sample app.
- Lastly, `cross-env` is a utility that allows us to write NPM scripts that work on shells across all operating systems.

Configuring NPM scripts

Next, let's add a set of scripts to our `package.json` file so as to automate running development and production builds. In that file, we will only change the `scripts` dictionary, replacing it with the following:

package.json (fragment)

```
"scripts": {
  "build": "cross-env NODE_ENV=production webpack",
  "dev": "webpack-dev-server --content-base public"
},
```

We'll explore these scripts and their purpose in the next section of this chapter.

Configuring Webpack

Lastly, we need to create a configuration file for Webpack. This is a fairly long document, so you can copy it from this book's code repository provided at the following link and paste its contents into `webpack.config.js`:

`https://bit.ly/sveltebook-webpack`

Let's examine the contents of the file.

After importing all the requisite modules, we define the `prod` variable, which is `true` if we're building the application for production, as determined from the NODE_ENV environmental variable:

```
const mode = process.env.NODE_ENV || 'development'
const prod = mode === 'production'
```

As we'll see in the next section, when building for production, we are enabling further optimizations, such as minification of the bundled JavaScript file. On the other hand, when building for development, we're enabling support for extra development tools, such as live reload and debugging.

The main part of the file is the configuration object for Webpack. This is a dictionary, and it's the exported symbol from the file:

```
module.exports = {
    entry: { /* ... */ },
    resolve: { /* ... */ },
    output: { /* ... */ },
    module: { /* ... */ },
    mode,
    plugins: [ /* ... */ ],
    devServer: { /* ... */ },
    devtool: prod ? false: 'source-map'
}
```

A lot of the content of the file is boilerplate code, but it's worth highlighting a few things:

- We define an entrypoint called `bundle`, starting from `src/main.js`.

- In the `resolve` dictionary, we are changing Webpack's defaults to better support Svelte modules downloaded from NPM, which often ship with uncompiled `.svelte` files.

- In the `output` configuration, we are telling Webpack to place the files in the `public` folder. Because the main entrypoint's name is `bundle`, the compiled JavaScript bundle will be located at `public/bundle.js`.

- The `module` dictionary ensures that Webpack loads Svelte files (with the `.svelte` extension) and handles them correctly, as well as `.css` files.

- In the `plugins` object we are loading the `dotenv` plugin, which we'll use later in the sample application to inject configuration values.

Webpack is highly customizable and features a lot of different options. It also comes with a large number of plugins. You can read more about configuring Webpack in the official documentation: `https://webpack.js.org/guides/`.

"Hello, Svelte!"

Now that we've scaffolded our project, we're ready to launch a "hello world" application!

This "hello world" application demonstrates that the Svelte project has been bootstrapped successfully, and that everything has been set up correctly.

Creating the App component

Let's start by creating our first Svelte component. Create the `src/App.svelte` file and paste the following content:

src/App.svelte

```
<h1>Hello, {name}!</h1>
<p>My first Svelte app</p>

<style>
p {
    color: #1d4585;
}
</style>

<script>
export let name = ''
</script>
```

Even before we get into the details of how Svelte components are written, which we'll do in the next chapter, you can already notice a few things.

To begin with, each Svelte component is placed in a file with the `.svelte` extension. Each file contains one, and only one, component.

The content of the Svelte component file looks like normal HTML pages:

- The layout is written using regular HTML tags.

- You can place CSS rules within a `<style></style>` block. All CSS rules defined here are scoped to the component. That is, the preceding rule only applies to p tags in the App component.

- JavaScript code is again put in a familiar place, between `<script></script>` tags.

In the first line, we're seeing for the first time the Svelte templating syntax: at runtime, `{name}` is bound to the value of the `name` variable.

The same `name` variable is also exported using the `export` statement, just like we'd do with ES2015 modules. The Svelte compiler looks for exported variables and makes them *props* (or *properties*), which can be set by outer components.

We'll cover components, the Svelte templating syntax, and props in much detail in the next chapters, but for now let's look at how we can initialize our App component.

The application's entrypoint

Svelte components can't be entrypoints for applications; only JavaScript files can be.

As you will recall, in the Webpack configurations, we defined the entrypoint as the `src/main.js` file. Let's create that:

src/main.js

```
import App from './App.svelte'
const app = new App({
    target: document.body,
    props: {
        name: 'Svelte'
    }
})
export default app
```

This short snippet imports the App component as a class. It instantiates the class, and finally exports the component as the application's code.

The constructor for the App class accepts a JavaScript object with a dictionary of options, including the following:

- `target` is the DOM element where the component should be rendered as a child; in our example, we want the application to be rendered in the page's <body>.

- The `props` dictionary is an optional one that allows a value to be passed to the component's props. Our App component is exporting a prop called `name`, and we're passing the string 'Svelte' as its initial value.

> **Note**
>
> In most cases, the app's entrypoint is the only place where Svelte components are manually instantiated with the new Component() syntax.

Index file

Lastly, we need to pre-create the index file that will load our Svelte application. Paste the following content into the `public/index.html` file (note the use of the public directory!):

public/index.html

```html
<!DOCTYPE html>
<html lang="en">
<head>
    <meta charset="utf-8">
    <meta http-equiv="X-UA-Compatible" content="IE=edge">
    <meta name="viewport" content="width=device-width,initial-
      scale=1">
    <title>Svelte Journal</title>
    <link rel="stylesheet" href="/bundle.css">
</head>

<body>
    <script async defer src="/bundle.js"></script>
</body>
</html>
```

This index file is a pretty standard HTML5 boilerplate page, but it does reference the /bundle.css and /bundle.js files, which are generated by the bundler.

Running the application

At the end of the setup, your project should look like the following screenshot:

Figure 2.3 – The "Hello, Svelte!" app

We're now ready to launch the application, using the `dev` NPM script:

- Open a terminal and run the following command:

```
$ npm run dev
```

- If you're using Visual Studio Code, you can also launch NPM scripts from the **NPM SCRIPTS** drawer in the bottom-left corner (if you just created a new project and you don't see it yet, relaunch the editor). Click on **dev** to launch it, highlighted in the preceding screenshot.

The bundler will compile your app and launch a local server. To see your application running, open a web browser and visit the following link:

```
http://localhost:5000/
```

Hello, Svelte!

My first Svelte app

Figure 2.4 – Our "Hello, Svelte!" application running

The script you launched is watching for code changes, and bundlers include live-reloading. Try making a code change (for example, change the text in the `App.svelte` component), and then save the file. Your browser will immediately reload to show your updates!

To stop the bundler from running and watching for code changes, press *Ctrl + C* in the terminal window:

> **Warning message**
>
> Running the preceding command, as well as the one in the next section, you'll likely see an error similar to `Failed to load ./.env` in the terminal window. This is just a warning that you can ignore for now; we'll create the `.env` file in the next chapter.

Compiling for production

Lastly, let's look at how the application can be compiled for production.

Running `npm run dev` compiles your application in a way that is designed for development. For example, it includes live-reload capabilities, it outputs source maps, and it does not minify or otherwise optimize the code for production.

To generate a production-ready bundle, run the following command:

```
$ npm run build
```

This build will take a few seconds more, but will minify all your JavaScript code and run other optimizations, including tree-shaking when possible.

After building for production, you will find your compiled, bundled application in the `public` folder, ready to be deployed (don't worry, we'll cover deployment options later in the book):

```
$ ls public
bundle.css
bundle.js
index.html
```

Debugging Svelte applications

Lastly, let's look at how we can debug Svelte applications in a web browser. There are two tools worth highlighting, which can be complementary: using the Visual Studio Code debugger, and using the *Svelte DevTools* browser extension.

Using the Visual Studio Code debugger

Visual Studio Code contains a powerful debugger that works neatly with web browsers such as Chrome, Firefox, and Edge.

Earlier in the chapter, I recommended installing the free **Debugger for Chrome** extension from the Visual Studio Code Marketplace (link to the extension in the Visual Studio Marketplace: `https://aka.ms/vscode-chrome-debugger`). That, or the equivalent for Firefox or Edge, plus a small configuration file is all we need.

Create a configuration file called `.vscode/launch.json` (note the dot at the beginning of the path) and paste the following content:

.vscode/launch.json

```
{
    "version": "0.2.0",
    "configurations": [
        {
            "type": "chrome",
            "request": "launch",
            "name": "Launch Chrome against dev server",
            "url": "http://localhost:5000",
            "webRoot": "${workspaceFolder}"
        }
```

```
        ]
    }
```

If you're familiar with the Visual Studio Code debugger, you'll notice that this is the standard, autogenerated file, with just the port changed in the URL.

In order to be able to set breakpoints in `.svelte` files, open the Visual Studio Code settings (**File | Preferences**, or **Code | Preferences** on macOS), search for `debug.allowBreakpointsEverywhere`, and then enable it.

After we've done this, we can use the full debugger, with support for breakpoints, data inspection, logpoints, and so on. To debug a Svelte application, perform the following steps:

1. First of all, start the dev server in a terminal; this is the same command we've run before:

    ```
    $ npm run dev
    ```

2. Secondly, click the **Run** icon in the Visual Studio Code activity bar to bring up the debugger and then press the green **RUN** button:

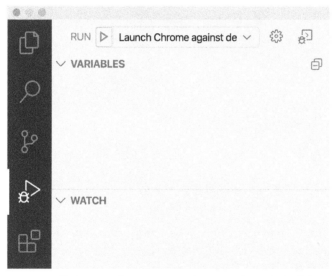

Figure 2.5 – Clicking on the green RUN button to launch the debugger

Alternatively, you can use the *F5* button as a keyboard shortcut.

3. Chrome will be launched automatically and will be connected to the debugger in the editor.

Visual Studio Code comes with a very powerful debugger, with support for advanced features, too. The best way to learn how to use the debugger is to check out the official documentation:

`https://code.visualstudio.com/docs/editor/debugging`

Browser extension

In addition to the debugger inside Visual Studio Code, this community-provided extension for Chrome and Firefox can be useful to inspect and change the state of your Svelte components and troubleshoot issues:

`https://github.com/RedHatter/svelte-devtools`

After installing the extension, when you're browsing a Svelte application running in a dev server (for example, our project launched with npm run dev), you will find an extra *Svelte* tab at the end in your browser's **Inspector** (also known as **Developer Tools**):

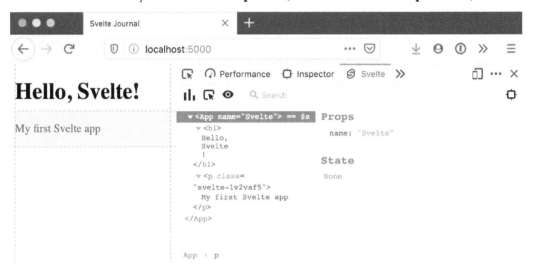

Figure 2.6 – The Svelte tab in the browser's developer tools

While not a full debugger, this extension shows the Svelte components that are currently rendered in the page, and their current state. You can change the values of props, for example, and see how the content changes.

Summary

In this second chapter, we've set up our development environment and scaffolded our project. We explored adding the Webpack bundler, and then we built our "hello world" application and got a first glimpse at how Svelte components are written. Lastly, we looked at debuggers and browser extensions to help us troubleshoot issues with our code.

In the next chapter, we'll dig deeper into this last topic; exploring the Svelte template syntax and other core features of the Svelte framework.

3
Building Reactive Svelte Components

Svelte is more than a framework: it's also a compiler, and, according to its creator Rich Harris, even a language. In fact, Svelte "extends" both HTML and JavaScript, in a natural way that has been applauded by developers.

In the previous chapter, we set up our environment and scaffolded the Svelte project. We also got a glimpse of how Svelte components work.

In this chapter, we'll be building the various Svelte components that our journaling app uses. Just as when learning any other programming language, using Svelte requires getting acquainted with a new syntax, which we'll explore gradually by building each component.

In this chapter, we'll be covering the following topics:

- Adding the required runtime dependencies and pre-made files that are specific to our journaling app

- Exploring the templating syntax for Svelte components and how to manage reactivity in components

- Using bindings and events, for regular HTML elements and components alike

Rather than listing all of the features of Svelte in a documentation-like fashion, the approach of this chapter and the next one is to present the concepts in a practical way. We'll introduce the syntax and the most relevant features of Svelte 3 as we encounter them in the components we'll be writing.

We believe that *Chapter 3*, *Building Reactive Svelte Components* and *Chapter 4*, *Putting Your App Together,* cover all of the most important (and most used) features of Svelte 3. They'll give you all the knowledge required to navigate the official documentation for the framework (which you can find at `https://svelte.dev/docs`) and further your knowledge using that if needed.

Adding requirements

Starting from the scaffolded project we created at the end of the previous chapter, we need to add a few modules from NPM that are specific to our journaling application. Additionally, we will need to add a pre-made script with functions that handle authentication and a style sheet with global rules.

Because these things are specific to our journaling application, they were not included in the scaffolding steps from the previous chapter, which were more generic for every application instead.

Launching the back-end service

In order for the following examples to work, you'll need to ensure that the API server is running on your laptop. Refer to the instructions in *Chapter 2*, *Scaffolding Your Svelte Project* for how to launch it.

Dependencies

Run the following command to add dependencies from NPM (because these modules are included in our application's bundle and not just used at build time, we are adding them as *dependencies* rather than *devDependencies*):

```
$ npm install --save \
    idtoken-verifier@2 \
    markdown-it@11 \
    svelte-calendar@1
```

Let's take a quick look at them, as follows:

- `idtoken-verifier` is used by the authentication library (which we'll add in just a moment) to validate **JSON Web Token** (**JWT**) session tokens.

- `markdown-it` converts Markdown into HTML: we use this so that our end users can write nicely formatted journal entries using Markdown (if you're not familiar with Markdown, check out `https://guides.github.com/features/mastering-markdown/`).

- `svelte-calendar` is a Svelte component that displays a datepicker; just like JavaScript libraries, Svelte components can be packaged and distributed using NPM.

Utilities

We'll also need three more files that you can copy from this chapter's repository on GitHub, found at `https://bit.ly/sveltebook-ch3`

Inside this `ch3` folder, look for these three files, and copy them in the same path in your local project:

- `src/lib/Session.js`—This file contains functions for dealing with session tokens. In particular, it receives session tokens passed by the authentication server (which, in our case, is inside the back-end service), validates the JWT session token, and extracts the *claims* it contains (for example, the user's name).

- `src/lib/Requests.js`—This contains functions that make requests to the back-end API server. They've been moved off the Svelte component files to keep this chapter's code smaller.

- `src/global.css`—This contains global styles for the application.

We won't get into the details of those scripts as they're outside of the scope of this book, and we'll just add them to our code as-is: consider them as any other external library! The code is thoroughly commented, so you're welcome to inspect it if you'd like.

The main.js file

Let's go back and revisit the `main.js` file that we created in the previous chapter, as we'll need to add to it code that uses the two files we've just downloaded.

Edit the file and replace its content with the following:

---

src/main.js

```
import './global.css'
import {HandleSession} from './lib/Session.js'
import {profile, token} from './stores'
import App from './App.svelte'

const app = (async function() {
    const [profileData, tokenData] = await HandleSession(0)
    profile.set(profileData || null)
    token.set(tokenData || null)

    return new App({
        target: document.body
    })
})()
export default app
```

As you can see, the changes are fairly minimal and are detailed as follows:

1. We've included `global.css` and `lib/Session.js` files. Webpack will deal with the style sheet and output a single CSS file automatically.

2. We're also importing a `store.js` file that contains the `profile` and `token` Svelte stores. We'll look into these in the next chapter.

3. We're encompassing the initialization code in a function that is immediately executed (in JavaScript, that's called **IIFE, or Immediately Invoked Function Expression**). We need to do that because our initialization steps include calling an asynchronous function (`HandleSession`) with the `await` statement, which requires us to be inside an `async` function and that can't be on the top level of the file.

4. Inside the initialization function we're invoking the code that sets up the session, and we're storing the two returned values (`profileData` and `tokenData`) into the two Svelte stores (again, we'll explore these in detail later).

5. Lastly, we're initializing our `App` component just like in *Chapter 2, Scaffolding Your Svelte Project,* but without props this time.

The dotenv file

We will need to create a *dotenv* file too. This is a file called .env (it starts with a dot) that contains a series of constants that are used by our app at build time. It's a convenient way to define parameters, such as where to find the back-end service with our APIs.

Create the following .env file in your project's folder (note that on Unix systems such as Linux and macOS, files that start with a dot are considered hidden files in the file explorer):

.env

```
AUTH_CLIENT_ID=00000000-0000-0000-0000-000000000000
API_URL=http://localhost:4343
AUTH_JWKS_URL=http://localhost:4343/jwks
AUTH_URL=http://localhost:4343/authorize?client_
    id={clientId}&response_type=id_token&redirect_
      uri={appUrl}&scope=openid%20profile&nonce={nonce}&response_
        mode=fragment
AUTH_ISSUER=http://svelte-poc-server
KEY_STORAGE_PREFIX=svelte-demo
```

Replace the value for AUTH_CLIENT_ID with a random **Universally Unique IDentifier (UUID)** to identify your application, which you can generate from https://www.uuidgenerator.net/version4. If you have multiple applications connecting to the same API back-end, this allows them to access different data.

The other variables contain mostly URLs: for the API server, the authorization endpoint, and the **JSON Web Key Set (JWKS)** used by the session management. The values are set to paths within http://localhost:4343, which assumes that your back-end server is running on your local machine. At the end of the book, when we go and deploy the code to production, we'll change them to point to an application running on a remote server.

The stores.js file

The main.js file imported another file called stores.js. Stores are a tool Svelte uses to share a state across components. Bear with us if this is unclear for now: we'll spend quite a bit of time talking about stores in the next chapter. For now, paste the following content in the src/stores.js file so that our app can compile:

src/stores.js

```
import {writable, derived} from 'svelte/store'
export const profile = writable(null)
export const token = writable(null)
export const isAuthenticated = derived([token, profile], (a) =>
  a && a[0] && a[1])
export const view = writable(null)
```

The index file

Lastly, we will need to make some small adjustments to the index file too, located in public/index.html, as follows:

1. Add these two <link> tags within the document's <head> and </head> tags to include two external CSS libraries: Tailwind CSS (a CSS framework) and fork-awesome (an icon set):

    ```
    <link rel="stylesheet" href="https://cdnjs.cloudflare.
      com/ajax/libs/tailwindcss/1.4.6/tailwind.min.css" />
    <link rel="stylesheet" href="https://cdnjs.cloudflare.
      com/ajax/libs/fork-awesome/1.1.7/css/fork-awesome.min.
        css" />
    ```

2. Edit the <body> tag to add some styles (these are powered by Tailwind CSS), as follows:

    ```
    <body class="bg-gray-100 text-gray-900 tracking-wider
      leading-normal">
    ```

Using Tailwind CSS

Tailwind CSS is a CSS framework for building web applications that look good and not "cookie-cutter" (like other frameworks), without too much effort. Almost all CSS classes used in the code come from it.

In this sample app, we're including Tailwind CSS precompiled from a **Content Delivery Network (CDN)**, but the recommended way would be to integrate it with your application's code as part of the build steps. This allows preprocessors to remove all unused CSS classes, greatly reducing the file size. Instructions for using Tailwind CSS with Webpack are on the official website: `https://tailwindcss.com/docs/installation`.

Svelte templates and reactivity

Let's start by creating our first components and explore how Svelte deals with templates and reactive statements.

Naming conventions

Names for Svelte components always begin with an uppercase letter. This makes it possible to distinguish Svelte components from HTML tags, which always start with a lowercase letter. For example, `Input` refers to a Svelte component, while `input` is the usual HTML tag.

Of particular importance when building a full web application with Svelte is that the root component is conventionally called `App`.

As for organizing your source files, in this book we'll be storing Svelte component files in the `src/components/` folder, in a file carrying the same name as the component. For example, the `App` component will be located in `src/components/App.svelte` (recall from *Chapter 2, Scaffolding Your Svelte Project,* how each `.svelte` file contains one—and only one—component).

Renderer.svelte component

The first component we're creating is the one that displays our journal entries and renders the Markdown syntax. We'll build it in multiple steps, placing it in the `src/components/Renderer.svelte` file.

This component accepts two props: the entry's title (a string) and content (Markdown, which will have to be rendered), as illustrated in the following code block:

src/components/Renderer.svelte (Step 1)

```
{#if title}
    <h1 class="text-3xl mb-3">{title}</h1>
{:else}
    <h1 class="text-3xl mb-3 text-gray-600">No title</h1>
{/if}
<div class="rendered">
    {#if content}
        {content}
    {:else}
        <p class="text-gray-600">No content</p>
    {/if}
</div>

<script>
export let title = ''
export let content = ''
</script>
```

This component looks similar to the one we saw in *Chapter 2, Scaffolding Your Svelte Project,* but we're starting to introduce some more complex templating syntax.

Let's start from the script at the bottom: as we've seen before, this component is exporting two props, `title` and `content`; both have an empty string as the default value. By exporting them as props, we are allowing other Svelte components to pass the values for them, as we'll do shortly ourselves.

Just as we've seen in previous chapter, the code outside of the `<script></script>` tags is regular HTML markup, with the addition of a templating syntax for Svelte.

Again, just as we've encountered before, the two blocks, `{title}` and `{content}`, instruct Svelte to render the values of the title and content in place, respectively.

The first new thing is the {#if val} ...markup1... {:else} ...markup2... {/if} block. As you would expect, this is a conditional statement: if the value of val is truthy (as per regular JavaScript expressions), it renders the markup in markup1; otherwise, it renders markup2. While not shown here, you can also use {:else if val2} to create additional conditions before an {:else} block.

Using the Renderer component

Congratulations on creating the first component of our journaling app! Now that you've created it, let's try to add it to our app.

For now, replace the content of src/App.svelte with this:

src/App.svelte (temporary)

```
<Renderer title="Hello" content="World!" />
<script>
import Renderer from './components/Renderer.svelte'
</script>
```

Then, start the development server (npm run dev) and open your browser to http://localhost:5000/ to see the page running. Note that if everything's configured correctly, you'll be first redirected to the auth page to authenticate yourself; use svelte as both username and password in the auth server. The Renderer component can be seen in the following screenshot:

Fig 3.1 – The Renderer component running

By doing this, you have seen something new: how to include a Svelte component in another component.

In the App component, in fact, we've imported the Renderer component with a normal ES2015 import statement, referencing its filename.

Then, we've added the component to our markup. You can reference a Svelte component in your markup by using it as if it were yet another HTML tag, anywhere in your markup (you can put it inside other HTML elements too, such as within `<div></div>` blocks, and so on). Note that unlike HTML tags, however, tags for Svelte components must always be closed, either explicitly (`<Renderer></Renderer>`) or self-closed (`<Renderer />`).

We've also passed the values for the `title` and `content` props, just as we'd pass attributes to HTML tags.

Adding Markdown rendering

The `Renderer` component we've just created is incomplete: in fact, you might have noticed that it does not do any rendering at all! In order to render the `content` prop from Markdown, we need to change it (changes are in bold), as follows:

src/components/Renderer.svelte (final)

```
{#if title}
    <h1 class="text-3xl mb-3">{title}</h1>
{:else}
    <h1 class="text-3xl mb-3 text-gray-600">No title</h1>
{/if}
<div class="rendered">
    {#if content}
        {@html rendered}
    {:else}
        <p class="text-gray-600">No content</p>
    {/if}
</div>

<script>
export let title = ''
export let content = ''
import MarkdownIt from 'markdown-it'
const markdown = new MarkdownIt()
$: rendered = markdown.render(content)
</script>
```

After importing the `markdown-it` library and initializing the `markdown` object, in the `<script>` component we're defining a `rendered` variable that contains the HTML rendered from the `content` prop.

The most interesting thing is the `$:` syntax, which makes the statement reactive. That semi-obscure syntax is actually a label in the JavaScript language, which Svelte adopts to tell the compiler to mark a statement as reactive. This is some of the magic of Svelte 3!

Reactive statements are executed every time the value of a variable changes. So, Svelte will re-render our Markdown every time the `content` prop changes and will put the result in the `rendered` variable. Without the `$:` label, Svelte would execute our statement only once, and if `content` changes, it won't update the rendered one.

Another thing to notice is that the `rendered` variable was not initialized, but Svelte will do that as if we had used the `let` keyword.

Lastly, in the markup, we've changed from `{content}` to `{@html rendered}`. Besides changing the name of the variable, we've added `@html` to indicate to Svelte that the `rendered` variable contains HTML tags that should be rendered by the browser. In fact, by default, Svelte encodes all HTML special characters (such as `<>"'&`) to prevent **Cross-Site Scripting (XSS)** attacks. Note that `@html` can be potentially dangerous, allowing XSS attacks, so make sure you trust the content you're trying to inject! (In our case, the HTML that the `markdown-it` library renders is supposed to be sanitized to strip any potential attack from malicious input.)

You can now update the `App` component and change the value of the `content` prop passed to `Renderer` to see how Markdown can be rendered, as follows:

src/App.svelte (temporary)

```
<Renderer title="Hello" content="{content}" />
<script>
import Renderer from './components/Renderer.svelte'
const content = `# Markdown here!
Here's a **great** post.`
</script>
```

In the first line, we're now telling Svelte to reference the `content` constant, defined in the script, for the prop with the same name (note how the value of `content` is delimited using backticks, `` ` ``, so we can define a multiline string whose line breaks are preserved by the JavaScript interpreter).

As you can see, the reference is defined by setting the value of the prop to {content}. The quotes around props are optional when passing an expression, so we could simplify that to the following:

```
<Renderer title="Hello" content={content} />
```

Additionally, when the prop uses a variable or constant with the same name, it can be simplified even further, like this:

```
<Renderer title="Hello" {content} />
```

While this last example might look odd at first as it's not valid HTML, the Svelte compiler expands it automatically by using content as the name for both the prop and the constant (or variable).

After making the code change, you'll now see the Markdown correctly rendered as HTML, as follows:

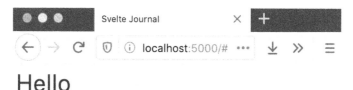

Fig 3.2 – Markdown is correctly rendered

Obj.svelte component

Let's create our second component, Obj.svelte. This component requests an object from the API server, then renders it using the Renderer component.

Awaiting on Promises

Paste the following code in src/components/Obj.svelte:

---

src/components/Obj.svelte

```
{#await contentPromise}
    Loading...
```

```
{:then response}
  <p class="mb-2 italic">
    Saved on
    {(response && response.date) ? new Date(response.date).
        toLocaleString() : '(null)'}
  </p>
  <Renderer title={response && response.title || ''}
      content={response && response.content || ''} />
{:catch err}
  <ErrorBox {err} />
{/await}
<script>
import Renderer from './Renderer.svelte'
import ErrorBox from './ErrorBox.svelte'
import {LoadObject} from '../lib/Requests.js'
import {token} from '../stores.js'

export let objectId = null
let contentPromise = Promise.resolve({})
$: contentPromise = LoadObject(objectId, $token)
</script>
```

The first new thing you'll notice in this component is the {#await} block in the first line. Svelte offers a built-in syntax in markup to *await* for Promises, such as those returned by async functions like LoadObject and built-in functions such as fetch, which is what's internally used by the former.

The syntax is: {#await promise} ...block1... {:then val} ...block2... {:catch err} ...block3... {/await}.

The flow is akin to using promise.then() and promise.catch() in JavaScript code, and is detailed as follows:

1. While the Promise named promise is in a pending state, block1 is rendered.

2. When promise is fulfilled, block2 is rendered, and it can use the Promise's final value in a variable named val.

3. If promise is rejected, Svelte renders block3, and the rejection reason is available in the variable named err.

Both the {:then} and {:catch} blocks are optional, and you can omit them if you don't care about the fulfilled value or the rejection reason.

The other thing you'll notice in this component, which should look relatively familiar now, is the usage of a reactive statement at the end of the script to assign a value to contentPromise. Every time the objectId prop changes, Svelte runs LoadObject and stores the Promise it returns into the contentPromise variable. The second argument to the function, $token, is a Svelte store, and as mentioned, we'll look into that in the next chapter.

Lastly, you'll notice that this component uses template expressions that are more complex than just variables or constant names. Here are some examples of this:

```
{(response && response.date) ? new Date(response.date).
   toLocaleString() : '(null)'}
{response && response.title || ''}
{response && response.content || ''}
```

response is the fulfilled value from the Promise and it's a dictionary that contains the date, title, and content properties.

Svelte template expressions can be full JavaScript expressions, including ternary conditionals (condition ? if-true : if-false), function invocations, and so on. For the preceding example, we're checking if response is truthy before accessing its properties to avoid exceptions when trying to access properties on non-objects.

Splitting into components: best practices

You might have noticed that the Obj component uses another component we haven't defined yet: ErrorBox. This is a rather small one, defined in src/components/ErrorBox.svelte, as follows:

src/components/ErrorBox.svelte

```
<div class="bg-orange-100 border-l-4 border-orange-500
   text-orange-700 p-4 my-2 mx-6" role="alert">
     <p class="font-bold">Error</p>
     <p>{err}</p>
</div>
<script>
export let err = ''
</script>
```

This component displays an error message and formats it. Because it's used by three different components that make network requests to display failures (Obj and two more we haven't created yet), it is a good practice to split out this code into a reusable component.

Trying the Obj component

To try the component(s) we just created, modify the src/App.svelte file, as follows:

src/App.svelte (temporary)

```
<Obj objectId="00000000-0000-0000-0000-000000000000" />
<script>
import Obj from './components/Obj.svelte'
</script>
```

Because there's no data stored by the API server yet, we're requesting the sample object with ID 00000000-0000-0000-0000-000000000000; the API server will return some content for our demo. Additionally, when using this sample object ID, the API server adds a delay of 3 seconds before responding, so you can see how the page is rendered when the Promise is in a pending state, as illustrated in the following screenshot:

Fig 3.3 – The Obj component showing the state when the Promise is pending and the sample document

Lastly, if you want to simulate the case when the Promise is rejected, simply shut down the API server, then refresh the page. Just as you'd expect from a JAMstack app (where **JAM** stands for **JavaScript, APIs, and Markup**), the front-end app will continue to appear, but it will be unable to retrieve the data from the back-end server and will show an error.

List.svelte component

We've explored the Svelte template syntax and learned about expressions, conditional statements, and using Promises. There's one more important element: loops.

Let's create a `src/components/List.svelte` component, which requests the list of journal entries and displays them. From the GitHub repository for this project, look in the `ch3` folder, then copy that file into the same location in your project's folder.

The component is very long, so we'll look at some highlights only. The component makes a request to the API server in the script and then waits for the Promise to be fulfilled: in that way, it's very similar to the `Obj` component we created moments ago.

The new concepts are in the markup, as follows:

src/components/List.svelte (fragment)

```
<!-- ... -->
<ul class="ml-6 space-y-2">
    {#each list as el}
        {#if el && el.oid && el.date}
            <li class="cursor-pointer bg-white shadow py-2 px-4
                w-2/3 lg:w-3/5"
            on:click={() => showObject(el.oid)}>
                <!-- ... -->
                <b>{el.title || '(no title)'}</b>
            </li>
        {/if}
    {/each}
</ul>
<!-- ... -->
```

First thing: yes, you can totally use HTML-like comments in Svelte markup! Those are then removed by the compiler.

As you can see, you can create a `{#each array as val, index}`...block...`{/each}` loop.

Just like using the `map` method on a JavaScript array, the `{#each}` statement iterates over the list and renders the `block` for each element. The markup inside `block` can access the current element of the array with the `val` variable and its index with `index`. Using `index` is optional, and, as you can see in the preceding example, it can be omitted if you don't need it.

There's one other new element in the preceding code snippet: an event when clicking on each `<li>` tag (`on:click={() => showObject(el.oid)}`). We'll look at events in the next section.

If you want to try the `List` component, you can add it to the `App` component as we've done with the other ones: just pass a Unix timestamp in seconds (that is, divide the result of `Date.now()` by 1,000) for the `date` prop. However, because there's nothing stored in the API server yet, you won't see any element returned.

Calendar.svelte component

This component shows a datepicker for users to select a day. From the GitHub repository, in the `ch3` folder, copy the `src/components/Calendar.svelte` file into your project, at the same path.

While this component is very important for our application's **User Experience** (**UX**), it is relatively less interesting from the point of view of what new we can learn from it. There are, however, two short snippets worth highlighting, as follows:

1. The first snippet shows that we can use a third-party component. You will recall how we installed the `svelte-calendar` component from NPM; this snippet adds it, as follows:

    ```
    <Datepicker bind:selected />
    <script>
    import Datepicker from 'svelte-calendar'
    let selected = /* ... */
    </script>
    ```

    As you can see, we import the component using a regular `import` statement, with the only difference being that the import path is the name of a module installed from NPM rather than the path to a file.

    This snippet highlights another feature: binding. This creates a two-way binding between a prop inside the `Datepicker` component and a local variable. We'll explore bindings in detail in the next section.

2. The second snippet is shown as follows:

```
<script>
/* ... */
$: {
    if (selected.getTime() != lastDate) {
        lastDate = selected.getTime()
        date = lastDate / 1000
    }
}
</script>
```

We've seen earlier how the $: label makes a statement reactive. If you need to make more than one line of code reactive, you can wrap that in a block and prefix it with $:, as you can see in the preceding code block.

Bindings and events

Let's work on another feature of the app: the form to add new content.

AddForm.svelte component

This is yet another long component, so I recommend copying and pasting it from the ch3 folder inside the GitHub repository. The path to the code is `src/components/AddForm.svelte`, and you should place that in a file at the same location in your project.

The AddForm component renders a form that, when submitted, invokes a handler that sends the data to the API server. The API server stores the journal entry and responds with a JSON object containing the ID of the newly created object.

Let's highlight some snippets from that file (most HTML tags and elements' class attributes have been removed for clarity), as follows:

---

src/components/AddForm.svelte (fragment)

```
<form on:submit|preventDefault={submit}>
    <input type="text" bind:value={title} />
    <textarea bind:value={content} class="... textarea-tall">
    </textarea>
    <button type="submit" disabled={running}>Save</button>
```

```
</form>

<style>
.textarea-tall {
    height: calc(75vh - 14em);
}
</style>

<script>
export let content
export let title
let running = false
</script>
```

The first thing worth noticing is that, among literally a dozen CSS classes applied to the `<textarea>` tag, one is defined in the `<style></style>` block below (the remaining 11 are from the Tailwind CSS framework).

As mentioned in *Chapter 2, Scaffolding Your Svelte Project,* you can define CSS classes in components, which are then applied by Svelte and scoped to the component only. After the code is built, Svelte will convert that CSS class to something like `.textarea-tall.svelte-kyz05b` that is applied to all tags with the `.textarea-tall` class in this component only (it is not applied in children components either, if present).

You can also see how the `running` boolean variable is used to control whether the `<button>` tag is disabled. Svelte knows that HTML attributes such as `disabled` are booleans, so it will cause them to be included when the expression is truthy, and they'll be omitted if the expression is falsey. That is: when `running` is `true`, Svelte renders the button as `<button disabled>`, while the `disabled` attribute is omitted if `running` is `false`.

All assignments to variables or properties of an object in Svelte are **reactive**. This means that if the script changes the value of a variable that is used in the template (just like the script in `AddForm` changes the value of `running`), Svelte automatically re-renders the component with the new state. Even better: because Svelte is compiled, it can intelligently re-render only the elements that have changed.

> **Reactivity and assignments**
>
> Reactivity applies to assignments to variables (`foo = 'val'`) or object properties (`obj.foo = 'val'`) only. Svelte does not automatically detect changes when you use a function that mutates the state of a variable internally (for example, using `push()` on an array).

There are two other new things that are worth analyzing: bindings and events.

Bindings on elements

Bindings allow data to "flow up", from an element, or—in another case—a component, to its parent component. The simplest and most common bindings are applied to input elements, which are then mapped bi-directionally to their value.

We've seen in the preceding snippet how we're binding the `title` variable to an input tag, as follows:

```
<input type="text" bind:value={title} />
```

At any time, accessing the value of the `title` variable in the script will return the current value of the input (as in its `value` attribute).

As mentioned, bindings are bi-directional, so assigning a new value to the `title` variable in the script will also change the value of the input as displayed to the user.

Events from elements

The other new concept from the preceding snippet is the **event** applied to the `form` tag, similarly to what we've seen in the previous section.

Using the `on:` keyword with an element, we can attach a handler to a specific **Document Object Model** (**DOM**) event that could be triggered by that element. An example can be seen in the following code snippet:

```
<button on:click={handler} />
<form on:submit={handler} />
```

The types of events you can listen to depend on what the HTML element is able to trigger; they are irrelevant to Svelte.

The handler is a function expression. It can be the name of a function defined in the script block, as with the `<form>` tag in the `AddForm` component, or an inline one, as we've seen done with the `List` component: `on:click={() => showObject(el.oid)}`. (Note how we're defining a new inline arrow function: using just `{showObject(el.oid)}` would not work, as that's a *function invocation* and not a *function expression*.)

Lastly, Svelte allows for modifiers to be added to events. These are shorthand ways of doing common tasks; the most common ones are the following:

- `preventDefault` runs the `event.preventDefault()` method before executing your handler (commonly used with `form` or a tags).

- `once` removes the handler after the event has been triggered the first time, so the handler runs only once.

- `self` runs the handler only if the target of the event is the element itself.

- `stopPropagation` runs `event.stopPropagation()` after the handler to stop the event bubbling to the next element.

You can add zero, one, or more modifiers, using | after the event's name. An example can be seen in the following code snippet:

```
<a on:click|preventDefault|stopPropagation={handler}>
```

Events from components

The last snippet worth highlighting from the `AddForm` component is the following one:

src/components/AddForm.svelte (fragment)

```
<script>
import {createEventDispatcher} from 'svelte'
const dispatch = createEventDispatcher()
async function submit() {
    /* ... */
    const objectId = await AddRequest(title, content, $token)
    dispatch('added', {objectId})
    /* ... */
}
</script>
```

The submit function is the handler for the on:submit event on the form tag, and we've removed all irrelevant lines.

This example is the first one in which we're importing a method from the Svelte runtime, as you see in the first line of the script. The createEventDispatcher method is used to create a dispatch function (second line in the script), which can be invoked in the script to trigger custom events **from the component**.

In the previous section, we looked at events triggered by elements, which are native DOM events. Just like those, component events are used to "bubble up" data, from a component to its parent. However, component events are created and triggered by a component's script and can have any name and data.

We'll see how to listen to custom events in the next section, as we write the parent component for AddForm.

ViewAdd.svelte component

The src/components/ViewAdd.svelte file contains the first view component, and it's the full layout of the page for adding new journal entries (our app's other views will be ViewList and ViewObject, which are used to list entries and read one, respectively). Create it by pasting this content:

src/components/ViewAdd.svelte

```
<AddForm on:added={added} bind:content bind:title />
<div class="bg-white shadow rounded px-8 pt-6 pb-8 mb-4
    text-sm">
    <h1 class="text-3xl mb-2">Preview</h1>
    <div class="border border-gray-600 p-4">
        <Renderer {title} {content} />
    </div>
</div>

<script>
import AddForm from './AddForm.svelte'
import Renderer from './Renderer.svelte'
import {view} from '../stores.js'

let content = ''
let title = ''
```

```
function added(event) {
    if (event && event.detail && event.detail.objectId) {
        $view = 'view/' + event.detail.objectId
    }
}
</script>
```

This is a relatively smaller component, but there are two points worth highlighting.

Bindings on components

The first point is how we're binding to a component's props. In the previous sections, we've seen how we can bind a variable to certain attributes of a DOM element (for example, when we bound a variable to a form input's value attribute).

You can also get a two-way binding to components' props.

If you recall, the AddForm component exported the content and title props. We saw at the beginning of this chapter how we can assign a value to a component's props, and this is illustrated in the following code snippet:

```
<AddForm content="foo" title="bar" />
```

We've also seen how the value can come from a JavaScript expression, such as content={content}.

Assignments to props such as these are one-way operations. When ViewAdd assigns a value to a prop of AddForm, the latter receives it and re-renders itself using the new value. The assignment can happen at any time, even after the component has been rendered already (this will trigger an update of AddForm). However, if AddForm were to change the value of the prop, it would only remain within the scope of that component, and ViewAdd would not get the updated value.

When using bindings, instead, there's two-way communication between the props of a component and a variable in its parent. This means that if AddForm changes the value of title internally, then the value of the title variable in ViewAdd changes too. You can create a two-way binding with the following code:

```
<AddForm bind:content={content} bind:title={title} />
```

When the name of the variable is the same as the name of the prop, you can shorten this to the following:

```
<AddForm bind:content bind:title />
```

Because we've created the binding, we can then pass `title` and `content` to the `Renderer` component too, as we've done in the fifth line in the preceding snippet.

In practice, the end result for our app is that we get a live preview of what the user writes in the form, with live Markdown rendering as they type! This is illustrated in the following screenshot:

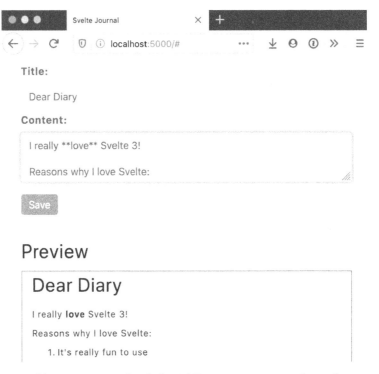

Fig 3.4 – The ViewAdd component rendered: the AddForm component at the top lets users type a new entry, and the Renderer component below it shows a real-time preview (with live Markdown rendering)

To render the `ViewAdd` component and try it yourself, write the following code in `src/App.svelte` (note that the component should be fully functional at this stage, so you can submit the form and have the API server store your journal entry!):

src/App.svelte (temporary)

```
<ViewAdd />
<script>
import ViewAdd from './components/ViewAdd.svelte'
</script>
```

Receiving events from components

The other new concept in the ViewAdd component relates to how it listens to the custom added event that the AddForm event triggers when an object has been saved in the server.

In the last snippet from AddForm written previously, you will remember this line, where objectId is a string with the ID of the newly created object:

```
dispatch('added', {objectId})
```

Note that the dispatch function's second argument must be a JavaScript dictionary.

This custom event can be received by ViewAdd with this code (irrelevant code, including bindings, has been removed for clarity):

src/components/ViewAdd.svelte (fragment)

```
<AddForm on:added={added} />
<script>
function added(event) {
    if (event && event.detail && event.detail.objectId) {
        // Do something with event.detail.objectId
    }
}
</script>
```

As you can see, the object we pass as the second argument for the dispatch function is available in the event listener as event.detail, so our objectId value is available at event.detail.objectId (all intermediate checks are to ensure that we don't try to access properties on nullish values, which would cause an exception).

Note that unlike events in the DOM, component events do not bubble up the component tree. If your component wants to pass a component event to its parent component, it needs to explicitly forward it. This is done by using the on: directive without a value. Here's an example of this:

```
<AddForm on:added />
```

The full added function written previously stores the object ID in a Svelte store called $view—as mentioned, we'll look into stores in the next chapter.

Summary

After fetching the required runtime libraries, we spent almost all of this chapter writing Svelte components for our journaling app. While doing that, we explored the syntax used by Svelte templates, and we dug into concepts such as binding and events, both for elements and for components alike.

Our app is not yet complete at this stage, as we're still missing a few components. To build them, which includes adding all the views and navigation between them, we need to learn a few more concepts first, including the long-promised Svelte stores, which we will cover in the next chapter.

4
Putting Your App Together

We ended *Chapter 3, Building Reactive Svelte Components,* with seven Svelte components created for our app. We were able to test them individually and see the pieces of our project coming together. We still need to create three more components, including the views that put everything together and make navigation possible between the various parts of the app.

While creating the last components, we'll learn about managing cross-component state using Svelte stores. We'll also look at using Svelte's built-in animations to make transitions nicer between views. Lastly, we'll look into a few more features of Svelte 3 that are useful to know for building your own apps, even though we could not use them in our Proof-of-Concept project.

In this chapter, you'll learn about the following topics:

- Managing cross-component state using Svelte stores.

- Adding transitions to components.

- We'll also lightly cover other useful features of Svelte 3 that are not used by our journaling app: slots, actions, life cycle methods, and meta elements.

In order to follow the next examples, make sure you've created all source code files that we've explored in *Chapter 3*, *Building Reactive Svelte Components*. You can also clone this book's repository from GitHub (`https://bit.ly/sveltebook`) and start from the code found in the `ch3` folder.

Managing cross-component state

In *Chapter 3*, *Building Reactive Svelte Components*, we looked at how to use props, bindings, and events to pass variables between components, including from parents to children and vice versa. However, sometimes it's useful to maintain a state that is shared across multiple components, even those not having direct relationships.

Svelte stores

Svelte stores can be used precisely for this purpose. They provide readable and writable stores that can be accessed by any component in the application, while being fully reactive. That means that if a component depends on a store and the value of such a store changes, Svelte automatically re-executes all reactive statements and templates that depend on it.

In the previous chapter, we created a `src/stores.js` file with pre-made content. Let's take another look at it here:

src/stores.js

```
import {writable, derived} from 'svelte/store'

export const profile = writable(null)
export const token = writable(null)
export const isAuthenticated = derived([token, profile], (a) =>
    a && a[0] && a[1])
export const view = writable(null)
```

In this file, we are importing two methods from the `svelte/store` module (part of the Svelte runtime library), and then we're defining four different stores: `profile`, `token`, `isAuthenticated`, and `view`.

Kinds of stores

There are three different kinds of stores that can be imported from `svelte/store`, listed as follows:

- **Writable** stores can have their value changed at any time by components or scripts.

- **Readable** stores' value cannot be set by scripts or components from "outside".

- **Derived** stores are read-only ones whose value is derived from one or more other stores.

In the preceding example, we're defining `profile`, `token`, and `view` as writable stores, with a default value of `null`. This is the syntax:

```
const mystore = writable(initialValue)
```

Additionally, we're creating a derived store, called `isAuthenticated`, whose value depends on the ones of `token` and `profile`. This is the syntax for creating a derived store:

```
const mystore = derived(originalStore, (newStoreVal) => {
    return derivedVal
})
```

The first argument is another Svelte store object: `originalStore`. The second argument is a callback that is executed every time the value of `originalStore`, which receives the updated value, changes. The function should return a new value for the derived store.

In our preceding code, the `isAuthenticated` store depends on both `token` and `profile`: we are passing as the first argument an array with both stores. The second argument, the callback, is a function that returns a truthy value if the updated values of both `token` and `profile` are truthy: `(a) => a && a[0] && a[1]` where a is an array, `a[0]` is the new value of `token`, and `a[1]` is the new value of `profile`.

In our project, we are not using any readable stores as we didn't have a need for them. Nevertheless, you can create a readable store with this syntax:

```
const mystore = readable(initialValue, (set) => {
    /* Code to execute when there is the first subscriber to
        this store */
    set(newValue)
    return () => {
        /* Code to be executed when there are no more
            subscribers to this store */
```

```
    }
  })
```

The first argument is an initial value for the store, just like for writable stores.

The second argument is a function that is executed when the number of subscribers to the store goes from zero to one, and it can be used to update the value of the store. The callback receives one argument, a set function that can be called with the new value.

Lastly, the callback can (optionally) return a function that is executed when there are no more subscribers to this store.

Let's look at an example: a readable store that returns the current time and is updated every second. We could use this in a Svelte component to show a live clock, for example. Have a look at the following code snippet:

```
const time = readable(null, (set) => {
    set(new Date())
    const interval = setInterval(() => {
        set(new Date())
    }, 1000)
    return () => {
        clearInterval(interval)
    }
})
```

The initial value is null: the time doesn't matter when no one is reading it. Then, as the first component subscribes to the time store, we update the value to the current time, and then start an interval to update the time every second (1,000 milliseconds). More components can subscribe to the store and get the always-updating time. When no more components are subscribed, the final callback is executed, invoking clearInterval to stop the interval updating the store.

Readable stores can be used for other purposes—for example, to periodically poll a server for some data and distribute the result to multiple components, and so on.

Using stores in Svelte components

Stores are objects with a subscribe method and, optionally, a set method.

Inside a Svelte component, in the template or in a script block, you can access the value of a store by prefixing it with $ (for example, if your store object is foo, you can access and set its value directly with $foo). The Svelte compiler takes care of creating the right subscriptions when you're reading the value or using the set method if you're writing in the store.

We'll see an example of this in the next section, within the context of the Navbar component.

Accessing stores in JavaScript files

This section is specific to those situations when you need to access stores to read or set their value in scripts that are not compiled by the Svelte compiler—for example, in files with the .js extension.

We've seen an example of this in the src/main.js file we created in *Chapter 3*, *Building Reactive Svelte Components*, shown again here (irrelevant code was omitted from this snippet):

src/main.js (fragment)

```
import {profile, token} from './stores'
const app = (async function() {
    const [profileData, tokenData] = await HandleSession(0)
    profile.set(profileData || null)
    token.set(tokenData || null)
    /* ... */
})()
```

You can programmatically set a value in a Svelte store with the mystore.set(newval) method, as you can see in the preceding code snippet. Note that we're not putting the dollar-sign prefix to the store in this case, as that's a syntax that requires the Svelte compiler and only works in .svelte files.

Accessing a value programmatically is less straightforward, and it requires a subscription to be created. We've not done this in our project, so let's see how to access the time store we created earlier in this section, to get the current time. Run the following code:

```
time.subscribe((val) => console.log(val))
```

The preceding code will print the current time in the debugger console every second.

Additionally, the `subscribe` method returns a function that we can use to unsubscribe from the store; for example, this will show the current time only for 10 seconds and then will stop, as illustrated in the following code snippet:

```
const unsubscribe = time.subscribe((val) => console.log(val))
setTimeout(unsubscribe, 10000)
```

If you wanted to read the value only once, you could use the `get` function, which internally creates a subscription, reads the value, and then cancels the subscription (so, if you plan on needing to access the value frequently, it's recommended to manage a continued subscription in your own code), as illustrated in the following code snippet:

```
import {get} from 'svelte/store'
console.log(get(time))
```

Navbar.svelte component

Just as the name suggests, the `Navbar` component displays the navbar for the application.

From this book's GitHub repository, go to the `ch4` folder and copy the `src/components/Navbar.svelte` file into your project's folder, at the same path.

The file contains a lot of markup and styling that makes it relatively long, so let's look at the most relevant parts, starting with the `<script>` block, as follows:

---

src/components/Navbar.svelte (fragment)

```
<script>
import {profile, isAuthenticated, view} from '../stores.js'
let active
$: {
    switch ($view) {
        case 'add':
            active = $view
            break

        default:
            active = 'home'
            break
    }
```

```
    }
    </script>
```

In the script block, first we import three stores: `profile`, `isAuthenticated`, and `view`.

`view` contains the name of the currently loaded view, and it's used with a "makeshift router" for our app on the frontend. In the preceding code, we are determining the link to mark as active in the navbar by checking the value of the `view` store. To do that, we're accessing its value in the `switch` statement using `$view`. Additionally, we've marked the block as reactive (`$: { … }`) so that the `active` variable is updated every time `$view` changes.

The other two stores are used in the preceding template (most markup has been removed for clarity), as follows:

src/components/Navbar.svelte (fragment)

```
<span on:click={() => $view = null}
    class="ml-4 cursor-pointer {(active == 'home') ? 'text-
        blue-600 underline' : ''}">
    Home
</span>
<span on:click={() => $view = 'add'}
    class="ml-4 cursor-pointer {(active == 'add') ? 'text-
        blue-600 underline' : ''}">
    Add
</span>
{#if $isAuthenticated}
    <img src={$profile.picture}
        class="ml-6 w-auto h-12 inline-block"
        alt="{$profile.name} picture"
        title={$profile.name} />
{/if}
```

You can use stores in template expressions too, so we are showing the user's profile picture (which is a claim extracted from the **JSON Web Token** (**JWT**) and saved in the `$profile` store) only if the value of the `$isAuthenticated` derived store is truthy.

The two `span` tags are used to navigate between the views. When users click on them, the *click* DOM event is captured, and an inline function sets the value of the `$view` store. Because `view` is a writable store, you can update its value by assigning to the `$view` variable (starting with the dollar sign); the Svelte compiler takes care of generating the code to modify the value in the object.

If you go back and check the code we wrote in *Chapter 3, Building Reactive Svelte Components* now, you can see that we've used a `$store`-like syntax multiple times already, both to read the value of a store and to set it, inside components' markup and scripts. An example can be found in the `List` component (presented in the *List.svelte component* section of the previous chapter), where the `added` function (an event handler) set the `$view` store with `$view = 'view/' + oid`. Additionally, a similar syntax was used in the `ViewAdd` component (in the section named *ViewAdd.svelte component* in the same chapter).

Completing the app

At this point, we're ready to complete the app and get something that's fully functional!

To do that, we need to create the last two components (the two remaining views) as well as an updated `App` component.

From the `ch4` folder in the GitHub repository, copy these two files in your project's folder, maintaining the same path:

- `src/components/ViewList.svelte`
- `src/components/ViewObject.svelte`

At this point, there shouldn't be anything in those files you are not familiar with, so going through them is left as a comprehension exercise to the reader.

For the final (for now) `src/App.svelte` file, write the following code:

src/App.svelte

```
<Navbar />
<div class="container mx-auto w-full lg:w-3/5 px-2 pt-2 mt-2">
    {#if $view == 'add'}
        <ViewAdd />
    {:else if $view && $view.startsWith('view/')}
        <ViewObject objectId={$view.substring(5)} />
    {:else}
```

```
        <ViewList />
    {/if}
</div>

<script>
import Navbar from './components/Navbar.svelte'
import ViewAdd from './components/ViewAdd.svelte'
import ViewObject from './components/ViewObject.svelte'
import ViewList from './components/ViewList.svelte'
import {view} from './stores.js'
</script>
```

The file includes the Navbar component as well as all three views, and then displays the correct view based on the value of the $view store. This "router" has some drawbacks, and we'll see in the next chapter how to improve the routing on our client-side apps.

For now, run npm run dev and enjoy your app, now fully functional!

You can see the app in the following screenshot:

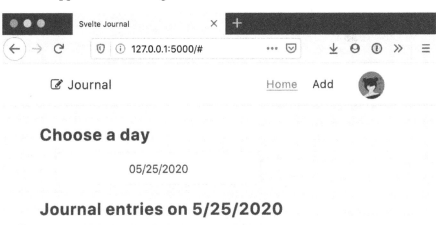

Figure 4.1 – Screenshot of the app with all the views in place

Transitions

Our app looks great already, but we could improve the user experience a bit with some transitions between the various views. With Svelte, this is available within the framework itself.

Svelte transitions

Svelte offers multiple pre-made transitions in the `svelte/transition` package. As of writing, there are six of them: `blur`, `draw`, `fade`, `fly`, and `scale`, `slide`. We will leave it as an exercise to the reader to try them and see their behavior!

> **Applying transitions**
>
> Transitions can only be applied to elements (that is, HTML tags) and not to Svelte components.

To make a DOM element transition nicely when appearing or disappearing, we just need to add a `transition:` property to its tag, followed by the name of the transition function. For example, take this snippet:

```
<button on:click={() => show = !show}>Click me</button>
{#if show}<p transition:blur>Hello world!</p>{/if}
<script>
import {blur} from 'svelte/transition'
let show = true
</script>
```

We've imported the `blur` transition from `svelte/transition`, and then applied it on the p tag with `<p transition:blur>`. Clicking on the button will cause the text to appear and disappear, and every change comes with a nice transition.

> **Svelte REPL**
>
> The Svelte project maintains an online REPL, or interactive shell, that lets you play with creating simple Svelte components within the browser itself. The REPL is available at `https://svelte.dev/repl`.
>
> You can even share snippets—for example, the preceding one is available at `https://bit.ly/sveltebook-repl-1`.

Using the `transition:` attribute applies that to transitions in both directions: in and out. To apply a transition only to one direction, use `in:` or `out:` respectively instead.

> **Tip**
> You can also combine the two to use different transitions for each direction—
> for example, `<p in:fade out:blur>`.

Additionally, transitions support parameters that can be used to tweak their behavior, such as their duration, and so on. Each transition function offers different parameters, so I will refer you to the official documentation to learn about them: `https://bit.ly/svelte-transitions`.

Adding transitions to our app

Let's add transitions to our app so that each view fades in nicely.

To do that, we will need to modify each of the three view components (`ViewAdd`, `ViewList`, `ViewObject`) in two ways, as follows:

1. Adding this `import` statement to the script block, like this:

    ```
    import {fade} from 'svelte/transition'
    ```

2. Wrapping all markup in a `div` tag with the `in:fade` attribute.

For example, the updated `ViewObject` component looks like this (code changes are in bold):

```
<div in:fade>
    <Obj {objectId} />
</div>
<script>
import {fade} from 'svelte/transition'
import Obj from './Obj.svelte'
export let objectId = null
</script>
```

Updating the other two components is an exercise we're leaving to the reader; you can also find the final code in the GitHub repository (in the `ch4-transitions` folder).

We're only setting a transition in one direction, when the component comes in the view, because otherwise we'd have a component transitioning out while the other one is transitioning in at the bottom of the page (I invite you to try it yourself to see the effect in action).

After making the changes to all three components, the app now shows smooth transitions when changing views.

Other useful Svelte features

We've tried designing our journaling app to make use of most of the features of Svelte 3 that you're likely to adopt when building your own apps, but we could not fit them all in. The rest of this chapter will lightly touch on some more features of the Svelte compiler and runtime that might be useful to you.

Slots

Slots allow parents to pass markup to child components.

In the child component, define the place for a slot with the `<slot>...</slot>` syntax; you can optionally put some default (fallback) content inside.

An example of this can be seen in the following code snippet:

Child.svelte

```
<h1>Here's the child component</h1>
<div style="margin: 0 2em;">
    <slot>
        <p style="color: #999">This is the default
            content </p>
    </slot>
</div>
```

Another component can import the `Child` component and set new content for the `<slot>` block, as illustrated in the following code snippet:

```
<script>
import Child from './Child.svelte'
</script>
<Child>
```

```
    <p>Some content will go in here!</p>
</Child>
```

You can try this yourself with this REPL:
`https://bit.ly/sveltebook-repl-5`.

You can have multiple slots using named ones. The following code snippet shows an example of this:

Child.svelte

```
<h1>Here's the child component</h1>
<div style="margin: 0 2em;">
    <slot name="body">
        <p style="color: #999">This is the default
            content</p>
    </slot>
</div>
<slot name="footer" />
```

And then, use it with this:

```
<script>
import Child from './Child.svelte'
</script>
<Child>
    <p slot="body">Some content will go in here!</p>
    <div slot="footer" style="font-size: 0.5rem;">All rights
        reserved</div>
</Child>
```

Again, you can try this yourself with this REPL:
`https://bit.ly/sveltebook-repl-6`.

Actions

With actions, we can attach functions to DOM elements that are automatically invoked when the component is created. They are attached to elements with the use: attribute.

Actions are functions that receive two parameters: the node they're used on (as a DOM node object), and an optional `parameters` argument. An example of this can be seen in the following code snippet:

```
<script>
function myAction(node, parameters) {
    console.log('The node', node, 'was mounted to the DOM')
    if (parameters) {
        console.log('The action myAction was invoked with
            parameters', parameters)
    }
}
</script>

<div use:myAction></div>
<div use:myAction={42}></div>
<div use:myAction={{foo: 'bar'}}></div>
```

Try it on the REPL: `https://bit.ly/sveltebook-repl-2`.

Additionally, actions can return a dictionary with up to two more methods: `destroy()`, which is executed when the element is removed, and `update(newVal)`, called every time parameters are updated (the function receives the new parameters as the first argument). Both are optional.

Actions can be very powerful when used to form a library, as they allow reusable logics to be defined that are applied to multiple elements. For example, let's look at an action that changes the text color of an element every time for 0.8 seconds the mouse hovers over it. This is defined in a custom library called `color.js` that exports the action, and then a Svelte component uses it, as illustrated in the following code snippet:

color.js

```
export function color(node, params) {
    const originalColor = node.style.color
    const handleMouse = () => {
        node.style.color = params
        setTimeout(() => {
            node.style.color = originalColor
        }, 800)
```

```
    }
    node.addEventListener('mouseenter', handleMouse)
    return {
        destroy() {
            node.removeEventListener('mouseenter', handleMouse)
        }
    }
}
```

Component.svelte

```
<script>
import {color} from './color.js'
</script>
<h1 use:color={'#9B2C2C'}>Hover me</h1>
```

When the user moves the mouse over the element to which the action is applied, the text color changes to the value passed in the parameter. We also return a `destroy()` method that removes the event listener when we're done.

Try this live on the REPL: `https://bit.ly/sveltebook-repl-3`.

Lifecycle methods

The Svelte runtime exports a few methods that can be used in your components' script blocks to perform actions at certain stages of a component's lifecycle. All of them can be imported from the `svelte` module itself at runtime.

beforeUpdate and afterUpdate

Just like their names suggest, `beforeUpdate` and `afterUpdate` accept callbacks that are executed right before a component is re-rendered and right after that, respectively, as illustrated in the following code snippet:

```
<script>
import {beforeUpdate, afterUpdate} from 'svelte'
beforeUpdate(() => {
    console.log('Component is about to be re-rendered')
})
afterUpdate(() => {
```

```
        console.log('Component was re-rendered')
})
</script>
```

Note that both `beforeUpdate` and `afterUpdate` are executed the first time the component is rendered too. Specifically, for `beforeUpdate`, this means that the method is executed the first time when nothing has been rendered to the DOM yet.

Accessing the DOM directly

Within a Svelte component, you could access the DOM directly—for example, using traditional methods such as `document.querySelector`, and so on. However, that is considered an anti-pattern, and whenever possible you should use Svelte-native methods (for example, reactive statements) to make any changes to the DOM instead. This will prevent unexpected behaviors and will help keep your code well organized.

Exceptions are situations such as actions, as we did previously, which are not defined in Svelte components and usually have no other way to alter the DOM.

---

onDestroy

The `onDestroy` method is invoked right after the component was unmounted (removed from the DOM). This is illustrated in the following code snippet:

```
<script>
import {onDestroy} from 'svelte'
onDestroy(() => {
    console.log('Component was unmounted')
})
</script>
```

You can use this lifecycle method to perform actions such as stopping timers/intervals or performing a cleanup of objects in external libraries, and so on.

> **About** onMount
>
> Just as Svelte has an onDestroy lifecycle method, it also offers an onMount one, which is executed right after a component is fully rendered the first time. However, unlike with other frameworks, Svelte lets you put initialization code directly in the component's script block, just like we've done with all of our components so far. Using onMount with Svelte is thus not usually necessary.
>
> The only exception is if you are using Svelte to do server-side rendering, because onMount does not run on the server and will instead be executed on the client after it receives the "pre-rendered" content. Because we're building a JAMstack app (where **JAM** stands for **JavaScript, APIs, and Markup**), server-side rendering is not in the scope of this book.

Meta-elements

Svelte offers a few "meta-elements" that can be useful in certain scenarios. The most relevant ones are the following:

<svelte:body> and <svelte:window>

Using <svelte:body> and <svelte:window>, it's possible to add listeners to events happening on the window and on the page's <body> element. Additionally, those events are automatically removed when the component that added them is unmounted.

For example, to do some cleanup before the page is printed by the user, we can listen to the beforeprint DOM event, as follows:

```
<svelte:window on:beforeprint={handler} />
```

Using <svelte:body> works the same way, but it lets you listen to events on the page's body rather than on the window.

Additionally, with <svelte:window> you can also bind local variables to certain properties, as follows:

- innerWidth and innerHeight (read-only)
- outerWidth and outerHeight (read-only)
- scrollX and scrollY (read-write)
- online, which is an alias for navigator.onLine (read-only)

An example of this is shown in the following code snippet:

```
<script>
let isOnline
</script>
<svelte:window bind:online={isOnline} />
<p>Are we online? {isOnline}</p>
```

Try it on the REPL: `https://bit.ly/sveltebook-repl-4` (then turn Wi-Fi off on your laptop for a moment to see the change).

<svelte:head>

Using `<svelte:head>`, you can add content within the `<head>` block of the page, which is removed automatically when the component is unmounted. For example, you could add a new style sheet, as follows:

```
<svelte:head>
    <link rel="stylesheet" href="extra-style.css">
</svelte:head>
```

Summary

In this chapter, we've completed our journaling application, which is now featuring all views, and it's functional. To achieve that, we've learned about Svelte stores and transitions.

Additionally, we've taken a look at some other features of the Svelte 3 language and runtime that we can use to build applications, even though they were not used by our proof-of-concept one.

However, we're not quite done yet. In the next chapter, we'll look at two ways to do routing on the client side, replacing our "makeshift router" that we're using now with better options (and we'll see why our current solution is not ideal). We'll also look at some additional things to set up in our Svelte app to make it more robust.

5
Single-Page Applications with Svelte

With *Chapter 4, Putting Your App Together,* behind us, we now have a fully-working application that lets us add journal entries, list all of the entries in a given day, and read them.

We've been building the application following the principles of the JAMstack, but we also built it as a **Single-Page Application** (**SPA**). This is an app that, once built by Webpack, contains all the views in a bundle with a single HTML file.

Because of that, we are left with one major change left to do to our app: adding (proper) client-side routing. We'll look at what that is and how we can implement it in this chapter.

In addition to that, we will also look at implementing some additional tooling to improve the quality of our code: setting up automated testing with **Nightwatch.js** (although we won't be writing tests at this time) and enabling linting.

In this chapter, we'll learn about the following:

- The two kinds of client-side routers for SPAs and when to choose each one
- Adding routing to our application with a hash-based router
- Setting up the environment for automated testing with Nightwatch.js
- Setting up linting for our code with ESLint

In order to follow the examples in this chapter, make sure you've created all the source code files that we explored in *Chapter 4, Putting Your App Together*. You can also clone this book's repository from GitHub (`https://bit.ly/sveltebook-ch4`) and start from the code found in the same `ch4` folder.

Routing on the browser

Backend developers are very familiar with the concept of routing in server-side applications: it allows the application (running on a web server) to render pages and content based on the URL the users requested.

For example, if our proof-of-concept application were a more traditional server-side one, we'd have to define at least the following routes:

- `GET /` would return the list of all posts for a given day (with a date picker).
- `GET /add` would render the form to add a new post.
- `POST /add` would receive the data for the new post and store it in the database, then redirect the user to read the post.
- `GET /view/{id}` would render the post with the given ID.

Because our app is a SPA, however, we are only shipping a single `index.html` file that contains all the different views. The state of the application controls what view is currently rendered.

Client-side routers are precisely the components that allow our SPA to render the correct view depending on the state, as well as handle navigation between views.

> **Multi-page JAMstack apps**
>
> As we mentioned in the first chapter, not all JAMstack apps are SPAs. If your application uses multiple pages, with one HTML file for each view, you do not need to implement client-side routing. In fact, those apps can navigate between views by requesting a different HTML page.
>
> Multi-page JAMstack apps are commonly generated by static site generators, such as Hugo, Jekyll, and so on, and they're very popular for blogging and building presentational websites.

Let's look at how to implement a router for our application, starting with the makeshift one that we implemented in the previous chapter.

Inspecting our makeshift router

To start, let's look at how our application has dealt with routing so far. In the previous chapters, we implemented a trivial "router" for our application using the Svelte $view store to determine the current view. In the App component, a series of {#if}...{:else if}...{:else}...{/if} statements in the template controlled the component to load. The relevant snippet was:

src/App.svelte (fragment)

```
{#if $view == 'add'}
    <ViewAdd />
{:else if $view && $view.startsWith('view/')}
    <ViewObject objectId={$view.substring(5)} />
{:else}
    <ViewList />
{/if}
<script>
import {view} from './stores.js'
</script>
```

Changing the view then required setting a new value in the $view store, for example as we did, for in the List component: every list item had an on:click event that triggered a function that changed the store, as in this snippet:

src/components/List.svelte (fragment)

```
{#each list as el}
  {#if el && el.oid && el.date}
    <li on:click={() => showObject(el.oid)}>...</li>
  {/if}
{/each}
<script>
import {view} from '../stores.js'
function showObject(oid) {
    $view = 'view/' + oid
}
</script>
```

This "makeshift router" is so bare-bones that it works in the sense that it allows changing the view that is currently displayed to the user. However, you can probably imagine how this is not scalable (our list of routes can easily become messy as we start adding more views).

There are also issues with the **User Experience** (**UX**) that you might have encountered yourself while testing the application, as follows:

- Users cannot navigate between pages using the browser's *back* and *forward* buttons. These are critical components of the UX for web applications. If you try pressing *back* yourself, you'll notice you're sent back to the authentication page, which is not in fact in the SPA.

- Hitting the *refresh* button in the browser will always send you back to the main view (the list of entries).

- Lastly, there is no way to directly access a specific view through its URL. For example, you cannot get a link to a single journal entry: something such as /object/{id}. This is needed to allow users to open pages in different tabs or to share links with others.

To solve these issues and give our users the best UX, we need to implement a more advanced router.

Svelte does not offer an official router component (at least not yet, as I'm writing this); however, there are multiple options maintained by the community. We'll explore a few in the next sections.

Two approaches to client-side routing

Routing on the client requires a solution that both matches the URL shown in the browser's bar with the view that is rendered and allows the application to trigger changes.

How to build a router is way beyond the scope of this book. However, we do need to look at the two different approaches used by routers to maintain the history state/stack and allow navigation backward and forward to understand which one is best to use:

- **The HTML5 History API**: As part of the first batch of HTML5 features, browsers started implementing APIs to allow applications to change the URL shown in the browser's bar and create *synthetic routes* (the main methods are `history.pushState` and `history.replaceState`). For example, applications can change the URL displayed to the user from `/home` to `/page/2` without causing the browser to make a new request to the server to try and retrieve `/page/2`.

- **Hash-based routing**: this is based on the fact that every change in the URL's fragment (the part after the # symbol) causes web browsers to create a new entry in the history stack. For example, if your app is a SPA that uses a single `index.html` file, you navigate between the different routes by adding a fragment to the URL – for example, `index.html#/list` and `index.html#/book/123`. Because `index.html` can usually be omitted, your URLs when using hash-based routing look like `http://example.com/#/book/123`.

Hash-based routing is older and has been used since the very first browser-based applications. The HTML5 History API is more recent, but has excellent browser support nevertheless (and works in Internet Explorer 10 too!).

In order to understand which one of the approaches to use, let's look at a side-by-side comparison:

The HTML5 History API	Hash-based routing
Introduced with early HTML5 (Internet Explorer 10+)	Has been available "forever"

Clean URLs:
`example.com/page/2`
	Fragments:
	`example.com/index.html#/page/2`
	`example.com/#/page/2`

Search engine-friendly	Not good for SEO
Requires server-side support	No server-side processing required
Best for public-facing websites	Best for Software-as-a-Service (SaaS) apps

Comparison of the HTML5 History API and hash-based routing

As you can see from the preceding table, using the HTML5 History API allows cleaner URLs that look like "regular" URLs. These are indexed easily by search engines (such as Google, Bing, and so on). Using these APIs is recommended for websites that are public-facing – for example, those that don't require authentication.

On the other hand, hash-based routing uses URLs that contain fragments that must not be removed, and because of that, they're poorly indexed by search engines. While this is not an issue when building SaaS-like applications – for example, those that don't display data to users who aren't authenticated (such as our journaling app!) – it can be a showstopper for websites that are open and public-facing. Despite hash-based routing being older and having many disadvantages, it's far from obsolete and is still widely used, including by popular services such as Google's Gmail.

The stickiest point is that when using the HTML5 History API, you get "synthesized routes," which do require support from the server. For example, if users visited your SPA at `https://example.com/` and then navigated to `https://example.com/page/2`, that URL would not actually exist in the server. That is because your app is a SPA, the only file in your server's document root is `/index.html`, and `/page/2` does not exist. Because of that, if users that are currently visiting `https://example.com/page/2` tried refreshing the page, opening the URL in another tab, or sharing the URL with others, they'd get a `404 Not Found` error.

In order to support those scenarios, you'd need your SPA to be hosted by a service that understands (or can be configured to understand) client-side synthetic routes and return your `index.html` file when the URL doesn't exist in the server. Services with support for this functionality do exist, and we'll look at some in the next chapter too, but you won't be able to serve your app from object storage services.

> **Routers that use the HTML5 History API**
>
> In this chapter, we won't be able to get into the details of implementing routing with the HTML5 History API. If that's a requirement for your app, you can look into components such as `svelte-routing` (`https://www.npmjs.com/package/svelte-routing`) or `yrv` (`https://www.npmjs.com/package/yrv`).

Routing with hash-based routing

For the first example, we'll implement routing with `svelte-spa-router` (`https://www.npmjs.com/package/svelte-spa-router`), which uses hash-based routing.

You can find the final code for this section in the project's GitHub repository in the `ch5` folder (for a direct link, go to `https://bit.ly/sveltebook-ch5`).

To start, install the module with the following command. We're saving it as a *dependency* and not a *devDependency* because it will be included in our application's bundle:

```
npm install --save svelte-spa-router@2
```

Defining the routes object

Using `svelte-spa-router` requires defining your routes in a dictionary (object) where the key is the path to match, which can optionally use parameters, and the value is a Svelte component object.

Create a file called `src/routes.js` that includes our routes dictionary:

src/routes.js

```
import ViewAdd from './components/ViewAdd.svelte'
import ViewObject from './components/ViewObject.svelte'
import ViewList from './components/ViewList.svelte'
import ViewNotFound from './components/ViewNotFound.svelte'

export default {
    '/': ViewList,
    '/add': ViewAdd,
    '/object/:objectId': ViewObject,
```

```
        '*': ViewNotFound
}
```

The file starts by importing all Svelte components that we want to use as routes.

The file also exports a dictionary that maps paths from the URL (technically, the fragment in the URL) to routes. These must start with / and can optionally include a "catch-all" route at the end:

- / (exact match) shows the list of all entries (the ViewList component).

- /add (exact match) shows the form to add a new entry (the ViewAdd component).

- /view/:objectId matches all routes that start with /view/ and then have a string parameter containing the ID of the object; it displays the ViewObject component.

- Lastly, there's a catch-all route, *, which shows the ViewNotFound component if no other route matches the current path (the catch-all route must always be defined last).

> **More options**
>
> The svelte-spa-router component offers more options for defining routes, including using regular expressions for matching. To learn more about using those, as well as to learn about other features of the routers that we won't cover in this chapter, check out the documentation for the project at https://github.com/ItalyPaleAle/svelte-spa-router

Updating the App component

To add the router, we need to modify the App component to include the router from svelte-spa-router and our routes dictionary. To do this, change the file to the following:

src/App.svelte

```
<Navbar />
<div class="container mx-auto w-full lg:w-3/5 px-2 pt-2 mt-2">
    <Router {routes} />
</div>
<script>
import Router from 'svelte-spa-router'
```

```
import routes from './routes.js'
import Navbar from './components/Navbar.svelte'
</script>
```

If you compare the preceding file with the previous version, you can see we have replaced all the `View*` components with a single `Router` one, to which we're passing our `routes` dictionary. This last object contains all of our routes too, so we've removed from this file all the `import` statements that were directly importing routes.

With this single action, we've taken care of one of the two sides of routing: determining the view currently requested by the user based on the URL, then displaying it. The other side is navigating between the different views, and we'll look at that in the next sections.

The ViewNotFound.svelte component

In the route definition object, we have added a new component that will be shown when no other route matches. We need to create the `src/components/ViewNotFound.svelte` file:

---

src/components/ViewNotFound.svelte

```
<h1 class="text-xl">Not found</h1>
<p>This page does not exist.</p>
<p><a href="#/" class="text-blue-600 underline">View all
  entries</a></p>
<p><a href="/add" use:link class="text-blue-600 underline">
  Add a new one</a></p>
<script>
import {link} from 'svelte-spa-router'
</script>
```

This component shows a simple page to inform visitors that the requested route does not exist. More interestingly, however, it shows the two different ways we can trigger a navigation declaratively:

- You can create a normal link, but with the `href` attribute beginning with `#/`: `<a href="#/">View all entries</a>` is just a link to the `/` view.

- Alternatively, you can create a link to the view starting with `/`, and then add the `use:link` action: `<a href="/add" use:link>Add a new one</a>` will lead people to the `/add` view (the `link` action was imported from the `svelte-spa-router` module).

These are the two ways you can navigate between pages declaratively. There are other ways to trigger navigation actions programmatically, which we'll look at shortly.

Updating the ViewAdd component

Let's update another file – `src/components/ViewAdd.svelte` – and copy the following code in the **script block only**:

src/components/ViewAdd.svelte (script block)

```
<script>
import {fade} from 'svelte/transition'
import {push} from 'svelte-spa-router'
import AddForm from './AddForm.svelte'
import Renderer from './Renderer.svelte'
let content = ''
let title = ''
function added(event) {
    if (event && event.detail && event.detail.objectId) {
        push('/view/' + event.detail.objectId)
    }
}
}
</script>
```

This shows how we can trigger navigation programmatically. While before we were setting a new view by changing the value of the `$view` store, now we're using the `push` method from the `svelte-spa-router` module. This tells the browser to load a new page, triggering navigation forward.

> **Other navigation methods**
>
> In addition to the `push` method, you can import two more methods from the `svelte-spa-router` module to programmatically control navigation between pages: `pop` makes the user go back one page (equivalent to pressing the browser's *back* button) and `replace` changes the current page without adding a new entry to the user's history stack (similar to a redirect).

Updating the List component

The `src/components/List.svelte` file needs to be updated too. This file is fairly long, so you can copy and paste it from this book's GitHub repository, from the ch5 folder. The most significant (but not the only) difference is in this fragment, which starts as the following:

```
<li class="..." on:click={() => showObject(el.oid)}>
```

It changes to the following:

```
<a class="..." href={'#/view/' + el.oid}>
```

Aside from the small changes in the list of CSS classes (omitted here for clarity), we have removed the on:click callback that invoked the function that updated the $view store to trigger the navigation. Instead, we've replaced the element with an a tag that has a href attribute made by concatenating #/view/ with the element's object ID (el.oid).

When this code is executed, the browser links will be rendered like this:

```
<a class="..." href="#/view/6c360a6d-6f5e-4cf9-bffb-
   37f122ee15c9">
```

URL parameters – updating the ViewObject component

In the route definition object, we created a route for /view/:objectId and in the previous section, we showed how we built a link to that view.

objectId is a parameter from the URL (technically, from the fragment of the URL), which is extracted by the router. The router passes all parameters to components through the exported params prop.

Let's look at an example, by updating the src/components/ViewObject.svelte file:

src/components/ViewObject.svelte

```
<div in:fade>
    <Obj objectId={params.objectId} />
</div>
<script>
```

```
import {fade} from 'svelte/transition'
import Obj from './Obj.svelte'
export let params = {}
</script>
```

There are two changes, both highlighted in the preceding code block. Starting from the bottom, you can see that the component now exports a `params` prop, which is an empty object. The router passes all parameters extracted from the URL to the component using this exported prop.

Because our route contained a parameter named `:objectId`, the component can access that as `params.objectId`, just like we're doing in the second line of the preceding example.

Updating the Navbar.svelte component

We have one last component to update: `Navbar`. We will need to change the contents of the `src/components/Navbar.svelte` file; again, this is a fairly long document, so please copy and paste the new contents from the book's GitHub repository, looking at the file in the `ch5` folder.

The main change in this component is how we determine the link that is currently active to highlight its appearance (changing its color and adding an underline). Let's look at this fragment (most of the markup has been removed for clarity):

src/components/Navbar.svelte (fragment)

```
<a class="ml-4" use:active={{className: 'text-blue-600
  underline'}} href="#/">Home</a>
<a class="ml-4" use:active={{className: 'text-blue-600
  underline'}} href="#/add">Add</a>
<script>
import active from 'svelte-spa-router/active'
</script>
```

We're using another action, `active`, which is imported from `svelte-spa-router/active`. Using this action, the router automatically determines whether a link is active based on the value of its `href` attribute and adds a CSS class (or a list of classes) to it.

Note that in the preceding example, `{{ ... }}` is not new syntax: it's just a JavaScript object defined inside a Svelte template expression (in this case, we're passing an object with the `className` property to the `use:active` action).

Navigating around the app

At this point, we have updated all the code, and our app should be running with the new router. Launch the local server (npm run dev) and navigate to http://localhost:5000 to see the app running:

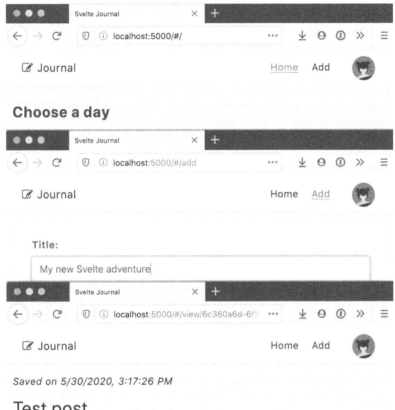

Figure 5.1 – Screenshot of the app running, showing how the different URLs show different views, and how the active link in the navbar is highlighted

As you can see by testing the app yourself or from the preceding screenshots, each page has a different path inside the URLs' fragment. In the navbar, the link that's active is highlighted in blue.

You can navigate back and forth using the browser's buttons. Additionally, refreshing the page will work too, and you can open links in new tabs as well.

At this stage, with the new router implemented, our app's code is complete! For the rest of this chapter (and book), we'll look at how to bring the app to production. We'll start by implementing automated testing.

Automated testing for Svelte apps

Apps are rarely "done," and they instead continue to evolve over time to respond to the needs of your users. One of the best ways to allow fast iterations while ensuring that your apps remain stable and with few bugs is to implement DevOps practices.

What is DevOps?

There are many definitions of what DevOps is, but one of my favorites is: "*DevOps is the union of people, process and technology to enable continuous delivery of value to our end users.*"

Embracing DevOps means adopting a culture and a set of processes to enable the development and operations teams to work together to deliver more value to users quicker. This includes adopting agile development practices, quick iterations based on feedback, open collaboration, and so on. Technology offers tools that can aid the adoption of DevOps, and we'll see some of them in this chapter and the following.

To learn more about DevOps practices and principles, you can refer to `https://azure.microsoft.com/overview/what-is-devops/`.

One of the pillars of DevOps is **Continuous Integration** (**CI**), which is made possible by automated testing.

In this section, we'll look at an introduction to how you can implement automated testing for Svelte applications using the **Nightwatch.js** framework. This is a free and open source tool that can be used to run CI testing for frontend applications.

Nightwatch.js allows controlling a web browser, such as Chrome, in headless mode (for example, without actually showing anything on screen) and lets you write tests to perform operations on web pages, such as clicking on links, submitting forms, and so on, programmatically, and then assert for the desired results.

This framework is especially useful for performing functional testing, ensuring that all components behave correctly. Of course, there are aspects of frontend application testing that can't be fully automated (for example, you could automate checking that a CSS class was added to a button, but humans will still be required to confirm that the page *looks good*). Despite that, being able to automate tests that confirm all components behave correctly can help to make development much faster.

Lastly, we should note that this section is to be considered a high-level introduction. We will look at enabling automated testing, but we won't look at how you can actually write your tests. That's a very large topic that could warrant an entire book by itself! Instead, you can find the documentation for writing tests with Nightwatch.js at `https://nightwatchjs.org/guide`.

Setting up the test environment

To start, we need to install Nightwatch.js from NPM, as well as at least one browser driver. In this example, we'll only test using Google Chrome, but you can find drivers for other browsers, such as Mozilla Firefox, Microsoft Edge, and Apple Safari.

Run the following:

```
npm install --save-dev \
    chromedriver@XX \
    nightwatch@1 \
    serve@11
```

Note XX used for the version of `chromedriver`: replace that with the version of Google Chrome running on your laptop (or, eventually, in your CI server). As of the time of writing, the CI platform we'll be using (GitHub Actions) includes version 84 of the browser (for GitHub Actions and Azure Pipelines, you can find the version of Chrome available in the base images at `https://aka.ms/actions-image-1804`).

In addition to Nightwatch.js and ChromeDriver, we're also installing `serve`, which is a small static server: we'll need it to create a local server for our compiled application. While the server that is started by `npm run dev` is great for development, to command generates a development build, which has some differences from the production one. Tests need to be run against production builds, so we first generate one with `npm run build`, and then we use `serve` to launch a local server.

Next, we need to add two new scripts to the `package.json` file. Inside, change the `scripts` block by adding two new lines (the changes are highlighted in the following code block):

---

package.json (scripts block only)

```
"scripts": {
    "build": "cross-env NODE_ENV=production webpack",
    "dev": "webpack-dev-server --content-base public",
    "serve": "npx serve -n -l 5000 public",
```

```
"test": "npx nightwatch"
},
```

Lastly, we need to create a configuration file for Nightwatch.js that tells it to use Google Chrome and ChromeDriver. This file is called `nightwatch.conf.js`, and you can copy it from the book's GitHub repository, from the `ch5-tests` folder: place it in the root folder of your project.

Creating tests

All tests must be placed in files with the `.test.js` extension within the `test` folder in your project. Test files are executed in alphabetical order, so it's common to name files with a number at the beginning.

As mentioned earlier, we won't be getting into test creation in detail, as that's a lengthy topic that's unrelated to the main purpose of this book. Instead, we'll look at two test suites as an example.

Create the `test/01-sample.test.js` test file and write the following:

```
/* eslint-env mocha */
describe('sample test', function() {
    this.slow(2000)
    this.timeout(3000)
    it('page renders', (browser) => {
        browser
            .url('http://localhost:5000')
            .expect.element('body').to.be.present.before(1000)
        browser.end()
    })
})
```

This file contains a single test that tells the browser to load our app running at `http://localhost:500` (which will be running with `serve`) and then ensure that the body element appears. This "hello world"-like test won't help much in the real world, but it's a good example of how to set up the test infrastructure.

In this book's GitHub repository, you can find another sample test suite in the `test/02-add-object.test.js` file within the `ch5-tests` folder. That file contains a more advanced example of a real-world functional test suite: after ensuring that the user is authenticated, it navigates to the route for adding a new journal entry, and it submits the form. While we won't analyze these other tests in our book, we're sharing the file as an example for interested readers to explore on their own!

Running tests

We should be ready now to run our automated tests. In order to run tests, we need to do the following:

1. Build the application. As mentioned before, it's best to run tests against a final, production-like build, rather than the development one returned by `npm run dev`; instead, run the following:

    ```
    npm run build
    ```

2. Start the local server and keep it running in a terminal:

    ```
    npm run serve
    ```

    After running this, you should be able to go to `http://localhost:5000` and see the application running.

3. Lastly, while the local server is running, in **another terminal**, run the following:

    ```
    npm run test
    ```

You should see the results for our single test shortly after:

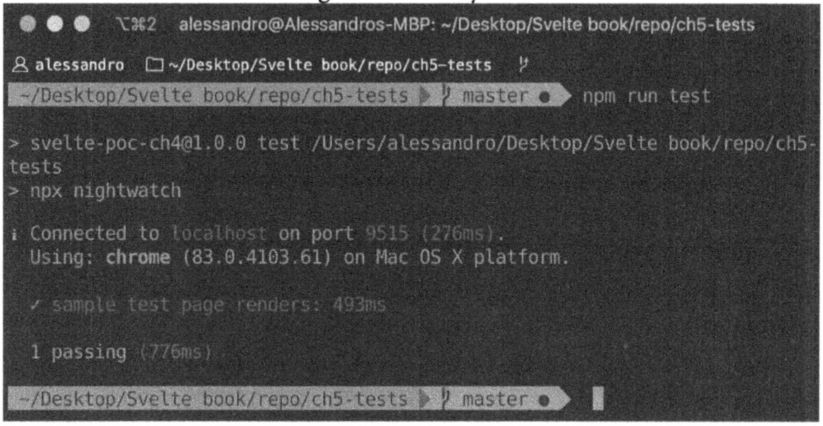

Figure 5.2 – Result of npm run test in the terminal

Linting and enforcing style conventions

Lastly, there's one more tool we will set up: **ESLint**. This is a linter, or in other words, a static code analyzer (that is, it inspects the code without running it) that is used for enforcing style conventions.

Tabs or spaces? Use semicolons or not? These are just two of the many style options developers have opinions about, sometimes really strong ones too! Regardless of what your preferences are, when you work in a team it's important to ensure that everyone writes code following the same styling conventions to keep the code base clean and readable.

ESLint is another open source tool that's almost ubiquitous among JavaScript developers, and it's a highly customizable linter. It allows you and your team to define rules for how the code should be written, and it can be used as part of your CI pipeline to enforce them. Even better, in many cases, ESLint can fix issues automatically, with the `--fix` switch!

Adding ESLint

Let's add ESLint and the modules for Svelte and HTML linting by running the following:

```
npm install --save-dev \
    eslint@7 \
    eslint-plugin-html@6 \
    eslint-plugin-svelte3@2
```

We also need to add a new script to the `package.json` file, adding one line to the `scripts` object (the new line is highlighted in the following code block):

package.json (scripts block only)

```
"scripts": {
    "build": "cross-env NODE_ENV=production webpack",
    "dev": "webpack-dev-server --content-base public",
    "serve": "npx serve -n -l 5000 public",
    "test": "npx nightwatch",
    "lint": "npx eslint -c .eslintrc.js -ext .js,.svelte,
        .html ."
},
```

We should also create a file called `.eslintignore` in our project's folder (note the dot at the beginning of the file's name) to tell ESLint to ignore files in the `public/` directory where we put our compiled application:

.eslintignore

```
public
```

Configuring ESLint for your style

Lastly, you need to configure ESLint's rules to match your desired style.

It's possible that your team already has its own ESLint rules that you normally use. In that case, you only need to make sure the following options are set in your `.eslintrc.js` file (note the dot at the beginning of the name) to be able to correctly lint code in Svelte files:

.eslintrc.js (fragment)

```
{
    env: {es6: true, node: true, browser: true},
    extends: 'eslint:recommended',
    parserOptions: {
        ecmaVersion: 2019,
        sourceType: 'module'
    },
    plugins: ['html', 'svelte3'],
    overrides: [
        {files: '**/*.svelte', processor: 'svelte3/svelte3'}
    ],
    settings: {
        html: {
            indent: 0,
            'report-bad-indent': 'warn',
            'html-extensions': ['.html']
        }
    }
}
```

In the preceding snippet, we are configuring ESLint to use ES2019 features, including ES modules (which Svelte relies on). We're then enabling the `html` and `svelte3` plugins, and telling ESLint to use `svelte3` for files that end in `.svelte`. Lastly, we're configuring the HTML linter too.

The preceding example uses the JavaScript file format for ESLint configuration with the `.eslintrc.js` file. It's also possible to configure ESLint using a JSON document (`.eslintrc.json`) or a YAML one (`.eslintrc.yaml`): if your team uses one of those formats, you might need to convert the rules from the preceding snippet into your desired format.

If you do not have an ESLint rule file, you can find the `.eslintrc.js` file I've used in this book's GitHub repository, inside the `ch5-tests` folder, which you can copy into your own project's folder.

Running the linter

Lastly, we can run the linter with the following:

```
npm run lint
```

The result will look as follows:

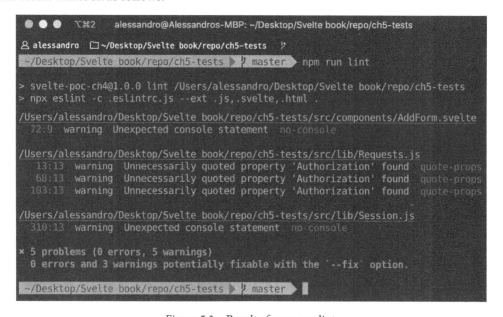

Figure 5.3 – Result of npm run lint

ESLint in Visual Studio Code

After installing and configuring ESLint, if you're using Visual Studio Code, you can see the results of the linter in real time within the editor.

In order to do that, first install the ESLint extension from the Visual Studio Marketplace at `https://aka.ms/vscode-eslint`.

After installing the extension, we need to configure it to run on both JavaScript and Svelte component files. Sadly, that option can't be changed using the **Settings** user interface yet, so we will need to manually edit the `settings.json` file in Visual Studio Code:

1. Open the command palette with *Ctrl + Shift + P* (or *Command + Shift + P* on a Mac), then type > `Preferences: Open Settings (JSON)` and press *Return*.

2. You'll see your Visual Studio Code settings as a JSON document. At the end of the file, add this element to the settings dictionary:

```
"eslint.validate": [
    "javascript",
    "svelte"
]
```

    Note that if you have other options in your `settings.json` file, you need to ensure that there's a comma before this option, as per the JSON specification.

After configuring the extension, you'll see the ESLint results automatically in Visual Studio Code while editing your code, in both JavaScript and Svelte component files.

ESLint errors and warnings appear as underlines in your code, and you can read the description of the issue by moving your mouse over them. Additionally, on the left sidebar, you can see the status for open files:

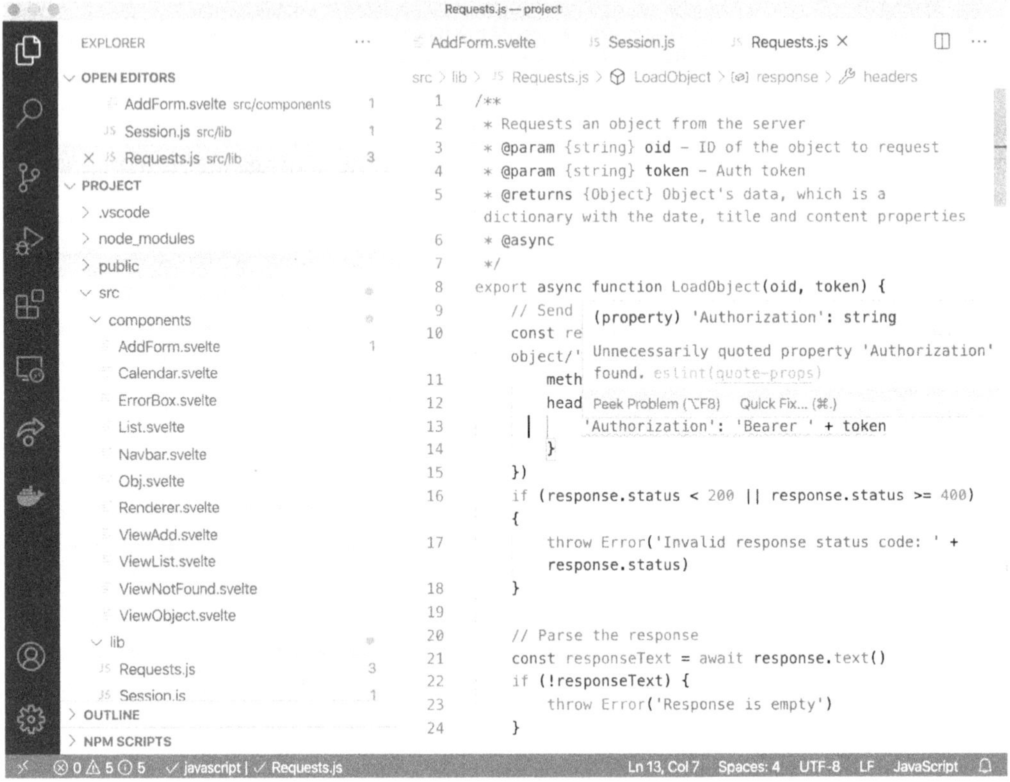

Figure 5.4 – ESLint support in Visual Studio Code, showing errors and warnings in the editor

Lastly, errors and warnings from ESLint appear in the **PROBLEMS** view, which you can summon by clicking on the warning and error symbols in the bottom-left corner on the status bar:

Figure 5.5 – Errors and warnings from ESLint in the PROBLEMS view

Summary

In this chapter, we completed our application by adding a client-side router (and we learned about the two different kinds of routers and when each one is recommended).

We also implemented two tools that can help us build applications more confidently.

First, we added the Nightwatch.js framework to enable automated testing using a headless browser. Now that the framework is in place and you know how to run tests, writing them can be a fun exercise that we'll leave up to you.

Lastly, we added ESLint to enable the linting of our JavaScript code and Svelte components, so we can ensure that everyone in our team writes code that looks consistent, and debates such as putting a newline between `if` and `{` can be put to an end (maybe!).

In the next chapter, we'll look at how to bring our application to production publicly. We'll also look at setting up continuous delivery to complete our DevOps pipelines.

6
Going to Production

Thanks to the work you've done in the previous chapters, our proof of concept journaling application is now code-complete. We were able to get it running on our laptops and generate production-ready bundles.

It's now time to actually deploy our application so that it can be accessed by others.

There are of course many, many ways to deploy apps, and your infrastructure can be as simple or as complex as you'd like. In this chapter, we'll explore a few options that are ideal for **JAMstack** applications such as ours (where **JAM** stands for **JavaScript, APIs, and Markup**), and we'll finally deploy our code using one of them.

In this chapter, we will cover the following topics:

- Learning about some of the ways Svelte apps (and JAMstack apps in general) can be deployed

- Deploying our application to an object storage service, Azure Storage, and enabling a **Content Delivery Network (CDN)**

- Setting up **Continuous Integration and Continuous Delivery (CI/CD)** with GitHub Actions

- Learning how to use the Azure portal to manage and destroy the resources we've created

This chapter requires a complete application, as with the one we built at the end of the last chapter. You can also find the complete code in this book's repository on GitHub (`https://bit.ly/sveltebook`) in the `ch5-tests` folder.

Options to deploy your Svelte apps

At the end of *Chapter 5, Single-Page Applications with Svelte,* our application was running locally on our laptops with the `npm run dev` command. Additionally, we were able to generate production-ready bundles with `npm run build` and start a local server from them. To make our application accessible to others, however, we need to deploy it to a cloud service.

Applications can be deployed to the cloud in a variety of ways. Cloud providers such as Microsoft Azure, Amazon Web Services (AWS), and Google Cloud offer a wide range of services that can be used to host applications, and you can architect very simple or very complex solutions depending on your needs.

The range of choice is especially large for JAMstack applications; because they are just static bundles, they can be served in dozens of ways.

As mentioned in the introduction to this book, one of the advantages of JAMstack applications is that they can be hosted by **object storage services** such as Azure Storage, AWS Simple Storage Service (S3), and Google Cloud Storage. When paired with a CDN, this architecture allows an efficient distribution of our frontend application to clients worldwide.

Static web apps can be cached very efficiently by the edge nodes of the CDN, so your app will be replicated worldwide and your users will access it from a location that is geographically close to them. As far as performance and availability go, this is as close to the gold standard as you can get. Additionally, a solution such as this is usually very inexpensive to operate.

We'll spend the rest of the chapter covering how to deploy your Svelte application to a solution based on object storage services, leveraging Microsoft Azure as the cloud platform.

Alternative services

In addition to our chosen solution, there are many other ways you can architect your infrastructure to deliver a JAMstack application. Some offer additional flexibility, such as the ability to offer the necessary back-end support for apps that leverage the HTML5 History API routing on the client side (as discussed in *Chapter 5, Single-Page Applications with Svelte*).

An incomplete list of alternative services includes the following:

- Serving the static files through a traditional web server running in a **Virtual Machine** (**VM**) or a container (also on platforms such as Kubernetes). For example, you could deploy an instance of nginx or another web server and set up a virtual host to serve your static files. The official documentation for nginx has extensive guidance on how to do this: `https://docs.nginx.com/nginx/admin-guide/web-server/serving-static-content/`.

  With the proper configuration, you could also enable support for synthetic routes and enable usage of the HTML5 History API.

  This solution allows the greatest flexibility. You could deploy your app on a VM or container running on an infrastructure you're already operating, alongside back-end services. You can also control every aspect of the application and infrastructure, including redirects, networking, firewalls, and so on.

  The downsides include the overhead of creating and maintaining VMs and/or containers. Depending on your availability needs, you might also need to maintain replicas of your application, keep them synchronized, and set up load balancers.

- To reduce some management overhead while still maintaining at least a degree of control, you can look into **Platform-as-a-Service** (PaaS) solutions. These are services in which a cloud provider manages the underlying infrastructure and VMs for you, and you can deploy your applications on top of that. They are not specific to JAMstack applications, and they support a variety of back-end stacks too.

  There is a really wide range of options in this category, including platforms that allow you to run Docker containers (so that you can package your static web app together with a web server such as nginx and create a Docker container image), such as Azure Kubernetes Service (AKS), Azure Container Instances, and Azure Web App for Containers; Amazon Elastic Container Service (ECS), Amazon Elastic Kubernetes Service (EKS), and AWS Fargate; Google Cloud's Google Kubernetes Engine (GKE), and Google Cloud's Google App Engine flexible environment.

  Other cloud providers offer similar services, either for individual containers or on a container orchestration platform such as Kubernetes. You still maintain a high level of control but are also required to maintain and "patch" your web server and base operating system inside the Docker container image.

Cloud providers also offer application platforms that support multiple stacks, including—of course—static web apps. Examples include Azure Web Apps, AWS Elastic Beanstalk, and Google App Engine. Services such as Heroku can be placed in this category as well. They offer less control over your app and infrastructure but they maintain the web server for you, so you don't need to patch anything.

- Lastly, in the last few years and months, a number of services have emerged that are specifically designed for JAMstack application hosting. The best examples in this category are Netlify and Azure Static Web Apps.

These are developer-centric services that allow only static web applications to be hosted and offer features that are specifically optimized for them, including redirects (needed for supporting client-side routing with the HTML5 History API). They differentiate themselves by offering integration with *serverless* platforms for adding server-side code, and they offer a built-in CI/CD platform.

GitHub Pages can be put in this category too, although it's currently more limited in its capabilities. As long as your app's source code is on a public repository, however, this is available for free.

Deploying to object storage

In this section, we'll be deploying our Svelte application to an object storage service. We'll be using Microsoft Azure—and, specifically, Azure Storage—for hosting the application, and Azure CDN for serving it globally.

The architecture looks like this:

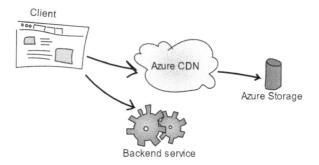

Figure 6.1 – Architecture diagram for the application

Creating an Azure account

If you don't have one already, you need to create an Azure account at https://azure.com/free.

You can create an Azure account by signing in using a Microsoft account (formerly known as a Windows Live ID) or a GitHub account.

Creating an account is free and should take just a few minutes. You will also be enrolled in a trial period with a certain amount of free credits that you can use within the first month. Additionally, for the first year, you can enjoy certain services for free: as of the time of writing, this includes 5 GB of storage in Azure Storage and 15 GB of bandwidth per month, which you'll be able to take advantage of.

Even after the free trial is over, your total Azure bill for hosting the solution should be really small.

Azure Storage costs depend on the amount of data you are storing, and it's billed at a few cents per GB per month. Static web apps per se are very small, normally in the order of a few KBs or MBs, but your storage needs might increase if you're serving a lot of large files, including videos or attachments.

As for the bandwidth, that depends on the amount of traffic your app gets and whether you're serving large files to users. On Azure, you pay for egress traffic only (that is, from Azure to your users, not the other way around), and it's billed at a few cents per GB used.

If you'd like to estimate your Azure costs, you can do so with the Azure pricing calculator, found at `https://azure.com/calculator`.

Authenticating with the Azure CLI

After you've created your account, you can manage your Azure resources in a variety of ways, including using the Azure portal and command-line tools. In this chapter, we'll primarily use command-line tools.

The Azure CLI works on Windows, macOS, and Linux. You can find installation instructions here: `https://aka.ms/installcli`.

After you've installed the Azure CLI, you can authenticate yourself by running the following command:

```
az login
```

This will automatically open your web browser pointing to the authentication page, where you can sign in with your Azure account.

Using Azure Cloud Shell

An alternative to installing the Azure CLI is to leverage Azure Cloud Shell, which offers access to the CLI within the web browser.

Azure Cloud Shell is available at `https://shell.azure.com/`. Sign in with your Azure account, then launch a **Bash** shell, as illustrated in the following screenshot:

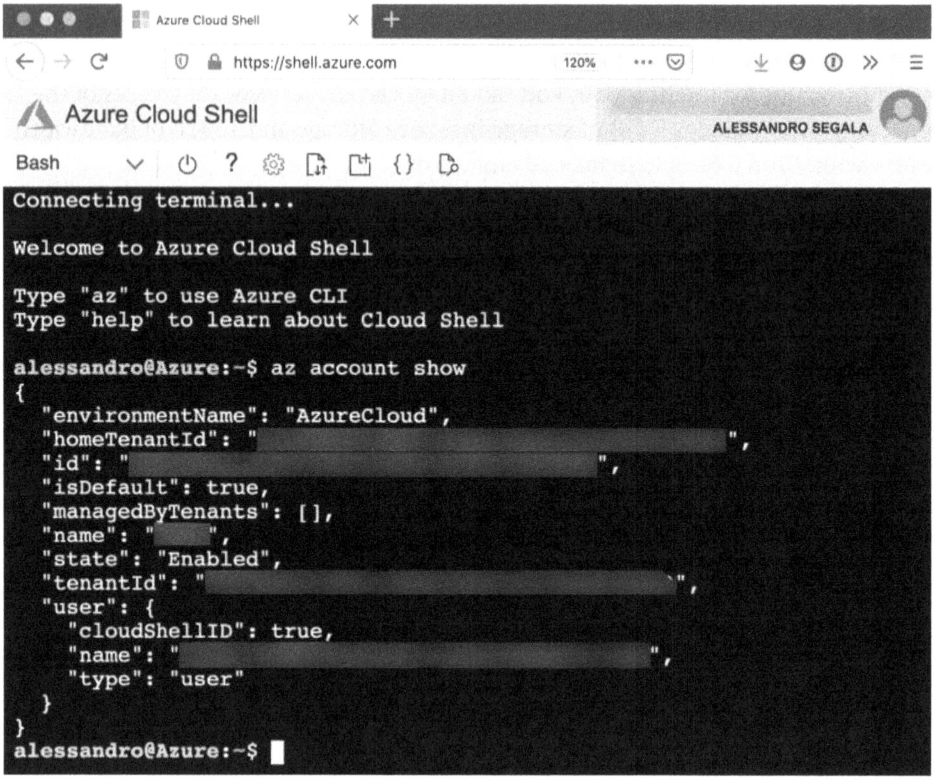

Fig 6.2 – Azure Cloud Shell

Using Azure Cloud Shell, you don't need to run the `az login` command to authenticate yourself again.

> **Using Cloud Shell in Visual Studio Code**
>
> Azure Cloud Shell is also available within Visual studio Code. After installing the "Azure Account" extension (`https://aka.ms/vscode-azure-account`), open the command palette (*Ctrl* + *Shift* + *P*; on macOS: *Command* + *Shift* + *P*) and type > `Azure: Open Bash in Cloud Shell`.

Creating the storage account

Now that we're ready to use Azure, we need to create the first of the two resources: an Azure storage account.

Azure Blob Storage is an object storage service where we can store files of every kind, either private or publicly accessible.

To start, we need to create a resource group, which is just a logical grouping unit for Azure resources. In the following code snippet, you might need to tweak the parameters in bold:

```
az group create --name Svelte-App --location eastus2
```

The --name parameter is the name of the resource group, and you can choose any name you'd like. The --location option is the name of the Azure region you want to use (you can find a full list with az account list-locations). Because our app will be served through a CDN, the choice of region to use is relatively less relevant. In fact, after the first user in each geographic location has requested your app, all other requests are cached by the CDN edge nodes.

Next, let's create an Azure storage account (again, parameters that you might need to tweak are in bold), as follows:

```
az storage account create \
  --name sveltejournal \
  --resource-group Svelte-App \
  --location eastus2 \
  --sku Standard_LRS \
  --kind StorageV2
```

The important options here are the following ones:

- --name is the name of the storage account. This has to be globally unique, so you'll need to choose a name that's unique to your app and deployment.

- --resource-group is the name of the resource group we created earlier.

- --location is again the name of the Azure region you want to use.

Lastly, we need to enable static website hosting, as follows:

```
az storage blob service-properties update \
  --static-website \
  --account-name sveltejournal \
  --404-document 404.html \
  --index-document index.html
```

--account-name is the name of the storage account that you created. The last two options are for defining the name of the files to show in case of a 404 error (page not found) and an index file (*note that the app we created did not have a page for 404 errors*).

Uploading files to your Azure storage account

Now that the storage account is created and configured, we can upload our app into a special storage container called $web, which was created automatically by the previous command.

There are multiple ways to upload files to a container inside Azure Storage.

If you're using the Azure CLI in your local machine, go to the folder where you have your Svelte app, and then run the following command. Make sure that you've bundled your app for production first, with npm run build! Once again, parameters that need to be tweaked are in bold in the following code snippet:

```
az storage blob upload-batch \
    --source public \
    --account-name sveltejournal \
    --destination '$web'
```

The arguments are listed as follows:

- --source is the path to the file or directory to upload; in this case, we're telling the CLI to upload all files in the public directory, which contains our compiled app.

- --account-name is the name of the storage account.

- --destination is the name of the container, which has to be $web for static web apps (don't forget to add single quotes around it so that the shell won't try to expand that as a variable).

There are other methods to upload or manage files inside Azure Storage. Here are a few popular options:

- Azure Storage Explorer, a cross-platform graphical application: https://azure.microsoft.com/features/storage-explorer/

- AzCopy CLI utility: https://aka.ms/azcopy

- Azure Storage extension for Visual Studio Code: https://aka.ms/vscode-azure-storage

Browsing the app from Azure Storage

In order to view a static website hosted in Azure Storage, we need to first retrieve its URL using this command:

```
az storage account show \
  --name sveltejournal \
  --resource-group Svelte-App \
  --query "primaryEndpoints.web" --output tsv
```

By now, you should be familiar with the `--name` and `--resource-group` parameters. The other parameters ask the command to print just the endpoint for the website.

The result will be something similar to this:

```
https://sveltejournal.z20.web.core.windows.net/
```

Here, `sveltejournal` is the name of your storage account, and `z20` is an internal identifier: both will be different in your case. The URL always ends in `web.core.windows.net`.

This URL isn't pretty. While it is possible to map a custom domain to a static web app in Azure Storage, it has some limitations (most importantly, no support for HTTPS). Because we'll be setting up a CDN in front of our app, we will map the custom domain to the CDN instead.

Adding the CDN

Adding the CDN via Azure CDN requires the following two commands to be run:

```
az cdn profile create \
  --location eastus2 \
  --name Svelte-App-CDN \
  --resource-group Svelte-App \
  --sku Standard_Microsoft

az cdn endpoint create \
  --profile-name Svelte-App-CDN \
  --resource-group Svelte-App \
  --name sveltejournalcdn \
  --origin sveltejournal.z20.web.core.windows.net \
```

```
--origin-host-header sveltejournal.z20.web.core.windows.net \
--enable-compression
```

As before, the parameters that need to be tweaked are in bold. They are the following:

- The location of the profile (`--location` in the first command): The CDN is global, but its metadata and configuration must be placed in a specific Azure region.

- The name of the profile (`--name` in the first command and `--profile-name` in the second) is a friendly name for our CDN profile.

- The resource group we created earlier (`--resource-group`).

- The name of the CDN endpoint (`--name` in the second command); this has to be globally unique, so choose one for your app and deployment.

- Lastly, we need to set the URL of the static web app in Azure Storage (`sveltejournal.z20.web.core.windows.net` in the example) for both the `--origin` and `--origin-host-header` flags in the second command.

If everything goes well, your CDN profile and endpoints will be created. Wait a few minutes, then you should be able to see your app running at the following address:

```
https://sveltejournalcdn.azureedge.net/
```

Here, `sveltejournalcdn` is the name of the CDN endpoint (the `--name` flag for the second command shown previously).

Mapping a custom domain

Most likely, you'll want to map a custom domain to your CDN endpoint so that your users can visit a URL such as `www.example.com`. Azure CDN supports custom domains and generates TLS certificates for them as well. This is a two-step process.

Mapping the DNS records

The first step is to update your **Domain Name System (DNS)** zone to create a CNAME record that points to your CDN endpoint.

The exact steps required to perform this operation depend on your domain registrar and/or your DNS service provider. We won't be able to provide instructions for all of them, so I recommend that you consult your provider's documentation for how to do this.

Generally speaking, you will map a sub-domain (such as `journal.italypaleale.me`) by creating a CNAME record for it that points to your CDN endpoint's hostname. In the preceding example, this is `sveltejournalcdn.azureedge.net`. You can choose any **Time-To-Live** (TTL) you want for your CNAME record, depending on your needs.

For example, when using Azure DNS as a DNS service, you could create the following record:

journal
italypaleale.me

🖫 Save   ✕ Discard   🗑 Delete   👤 Users   ⊘ Metadata

journal.italypaleale.me	🗗

Type

CNAME

Alias record set ⓘ

◯ Yes ⦿ No

TTL *   TTL unit

5	Minutes ∨

Alias

sveltejournalcdn.azureedge.net

Fig 6.3 – Example of creating a CNAME record for "journal.italypaleale.me" in Azure DNS

CNAME records and root domains

Per the DNS specifications, CNAME records can only be set on sub-domains. That is, while you can set a CNAME record on `www.example.com` and `journal.example.com`, you cannot set it on `example.com` (also called the "root domain").

If you want your CDN endpoint to be reachable on a root domain (and, by extension, your Svelte app), you need to use a DNS provider that supports "CNAME flattening,", sometimes called "ALIAS records." This is an advanced feature that is not available in all DNS providers.

If your DNS provider does not support that, you can consider switching to another service that does, such as Azure DNS: `https://docs.microsoft.com/en-us/azure/dns/dns-alias`.

Adding the domain to the CDN endpoint

The second—and last—step requires the configuration of your Azure CDN endpoint to accept your custom domain. This is done with the following two commands, to add the custom domain and enable HTTPS on it:

```
az cdn custom-domain create \
    --hostname journal.italypaleale.me \
    --endpoint-name sveltejournalcdn \
    --profile-name Svelte-App-CDN \
    --resource-group Svelte-App \
    --name sveltejournalcdn

az cdn custom-domain enable-https \
    --endpoint-name sveltejournalcdn \
    --profile-name Svelte-App-CDN \
    --resource-group Svelte-App \
    --name sveltejournalcdn
```

Most parameters are familiar. Pay attention to the name of the custom domain to add, which is the value for the --hostname flag in the first command and the value for --name in the second command.

Adding the custom domain and enabling HTTPS will take a few minutes. After this is done, you will be able to see the Svelte app running at your custom domain, as illustrated in the following screenshot:

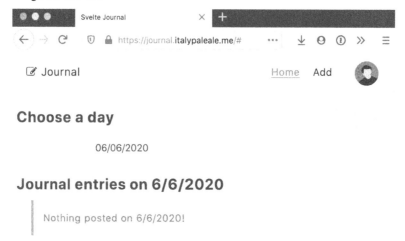

Fig 6.4 – Our app running on a CDN with a custom domain

The back-end service

If you tried navigating to your web app after one of the last steps (after it was deployed to Azure Storage, or after creating the CDN endpoint, or after adding the custom domain), you would have instantly been redirected to your back-end service to authenticate, running on your laptop. If the app wasn't running on your laptop, you would have seen an **Unable to connect** or **Can't reach this page** error in your browser (if a friend of yours tried it on their laptop, they'd have received the same error).

This is happening because, despite our app being deployed to the cloud, it's still configured to connect to the back-end API server on `localhost`.

I waited so long in this chapter to write about the back-end service precisely so that you could experience this: it is an excellent demonstration of a behavior that is unique to JAMstack apps. In traditional client-server applications, the web application frontend retrieves data stored in the web server; with a JAMstack app, the client can fetch data from any routable address.

Taking another look at *Figure 6.1*, depicting the architecture of the solution; you can see that the client is the one making connections to the back-end service. When the API service is running on your laptop and listening on `localhost:4343`, then your browser (and your web app) can request data from it. The back-end could be anywhere, and as long as your browser has a route to it, then you can fetch data from it.

Deploying the back-end service to the cloud

If you do want to make the back-end service public, you will need to deploy it somewhere.

A discussion on the different options to deploy a server-side application written in Go is outside of the scope of this book. If you want something quick, you can deploy it to Azure Web App for Containers. This is a platform service for running applications that are pre-packaged in Docker containers.

You can spin up the back-end app using the following commands, and then wait a few minutes for the deployment to complete:

```
az appservice plan create \
  --name Svelte-Backend-AppServicePlan \
  --resource-group Svelte-App \
  --is-linux \
  --location eastus2 \
  --sku B1
az webapp create \
```

```
--name sveltejournalbackend \
--resource-group Svelte-App \
--plan Svelte-Backend-AppServicePlan \
--deployment-container-image-name italypaleale/sveltebook
```

The preceding two commands create an App Service plan (a sort of managed web server) with the SKU "Basic 1" (the smallest one), and then create an application on top of that from the `italypaleale/sveltebook` container image from Docker Hub (the same one I recommended in *Chapter 2, Scaffolding Your Svelte Project,* as an option to run the app locally).

After the deployment is done, you can reach your app at the following address:

```
https://sveltejournalbackend.azurewebsites.net
```

Here, `sveltejournalbackend` is the value of the app's name: the `--name` flag in the preceding second command, which once again needs to be globally unique.

Important security notice

As mentioned in *Chapter 2, Scaffolding Your Svelte Project,* the API back-end service that we wrote was designed to be used as a companion for the development of our PoC app, but it should not be used for real, production solutions as is. To start, it neither offers the ability to register users nor to change the password for the default `svelte` user. By deploying the back-end to the cloud at a publicly accessible endpoint, you are open to abuse from people reading what you wrote and even writing their own content. Use this for testing only, and remember to shut it down at the end! (See the last section of this chapter for how to do this.)

Changing the configuration of the frontend

As mentioned, the frontend application is configured to communicate with a back-end application running on `http://localhost:4343`.

To point it to the app we deployed on Azure Web Apps, we need to change the content of the .env file we created in *Chapter 3, Building Reactive Svelte Components*, replacing all instances of http://localhost:4343 with the endpoint of our back-end running in the cloud. For example, see the following file, with changes in bold (remember to also set a custom **Universally Unique Identifier** (**UUID**) for AUTH_CLIENT_ID, as we did in *Chapter 3, Building Reactive Svelte Components*):

.env

```
AUTH_CLIENT_ID=00000000-0000-0000-0000-000000000000
API_URL=https://sveltejournalbackend.azurewebsites.net
AUTH_JWKS_URL=https://sveltejournalbackend.azurewebsites.net/
jwks
AUTH_URL=https://sveltejournalbackend.azurewebsites.net/
  authorize?client_id={clientId}&response_type=id_
    token&redirect_uri={appUrl}&scope=openid%20
      profile&nonce={nonce}&response_mode=fragment
AUTH_ISSUER=http://svelte-poc-server
KEY_STORAGE_PREFIX=svelte-demo
```

After making the preceding change, rebuild the static web app with npm run build, and then re-deploy it to Azure Storage, as you did earlier in the chapter, with the az storage blob upload-batch command or using apps like Azure Storage Explorer.

If you access your app through the CDN, you might need to purge the CDN's cache too, with this command (this can take up to 10 minutes):

```
az cdn endpoint purge \
  --profile-name Svelte-App-CDN \
  --resource-group Svelte-App \
  --name sveltejournalcdn \
  --content-paths '/*'
```

At this stage, our application is deployed to a cloud service and is running, and it's available to anyone in the world. This is a very important milestone already, and we could safely stop our work here.

For those who decide to follow along for the next bit, we will look at how to use DevOps practices such as CI/CD to automate the deployment steps, giving us more deployment agility and reducing the risk of errors by using automation.

Continuous Integration/Continuous Delivery

CI/CD are two very important aspects of DevOps. They consist of automatically building and running tests when a code change is committed to source control, and then automatically deploying the application (often to test environments before going into production).

We can enable full CI/CD for our Svelte application too, using a platform that runs our tests and automatically deploys to Azure Storage every time we make a code change.

We will use GitHub Actions as our CI/CD platform. This is built into the GitHub platform, and it's free for public repositories.

Putting our app's code on GitHub

We've reached this point without ever putting our app's code into any source control. Before we can run any CI/CD jobs, our app's code needs to be added to a Git repository and pushed to GitHub.

Installing Git

If you don't have Git installed already, follow these instructions, depending on your operating system:

- **Windows**: Download Git from the official website and install it: `https://git-scm.com/download` (if you're using the **Windows Subsystem for Linux** (**WSL**), follow the Linux instructions that follow).

- **macOS**: Git may already be preinstalled—for example, if you've installed Xcode or the Xcode Tools. If you don't have it, you can either download the official binary from `https://git-scm.com/download` or install it from Homebrew with `brew install git`.

- **Linux**: Git comes preinstalled in many Linux distributions. If it's not already installed in your system, you can fetch it from your distribution's repositories, with commands shown here: `https://git-scm.com/download/linux`.

After that, you need to set up Git by setting your username (your real name or a nickname) and an email address, as follows:

```
git config --global user.name "Your name here"
git config --global user.email "user@example.com"
```

Initializing a Git repository

The first step is to initialize a Git repository in your project's folder.

Using the terminal, run the following command:

```
git init
```

If you're using Visual Studio Code, you can also initialize the Git repository by opening the command palette (*Ctrl + Shift + P*, or *Command + Shift + P* on macOS) and typing > `Git: Initialize repository`.

Before we make our first commit, we need to create a `.gitignore` file (note the dot at the beginning) to exclude certain files from being committed into the repository. It's standard practice not to commit the `node_modules` folder (where NPM puts all installed modules), the compiled files (all files in the `public` folder except `index.html`), or any other documents used for debug or development.

You can find a `.gitignore` file in this book's GitHub repository (`https://bit.ly/sveltebook-ch6`) in the `ch6` folder. Copy it into your project's folder using the same name.

After that, we can make a commit importing all the existing code. Using the command line, run the following code:

```
git add "*" ".*"
git commit -m "First import"
```

Here, `"First import"` is the commit message, and it can be anything you'd like.

Using Visual Studio Code, you can do that from the **SOURCE CONTROL** tab (the third one from the top). First, click on the + next to **CHANGES** to stage all changes (hover your mouse over that row to make the button appear). This is illustrated in the following screenshot:

Fig 6.5 – The SOURCE CONTROL tab in Visual Studio Code

Write a commit message (for example, `First import`), then click on the checkmark above the message box to create the commit.

Pushing the code to GitHub

Next, we need to publish our repository on GitHub.

To start, you'll need a GitHub account if you don't have one already. You can get one for free at `https://github.com/`.

After signing in, create a new repository. In the top-right corner, you can click on the + symbol and then select **New repository**, as illustrated in the following screenshot:

Fig 6.6 – The New repository link to create a new repository on GitHub

Give your repository a name, add an optional description, then leave all other options as default and click on **Create repository**, as follows:

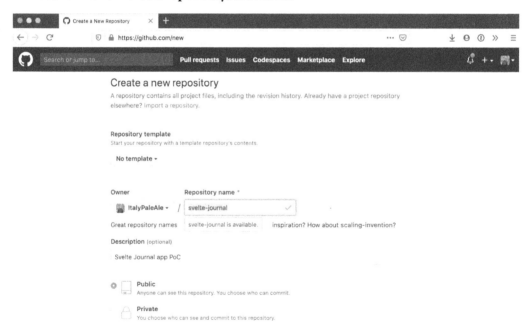

Fig 6.7 – Creating a new repository

The page that appears after you've created your repository already gives you the commands to run for the next step in the section titled **…or push an existing repository from the command line**. When choosing to use SSH, they look similar to these (the text in bold will be different for you):

```
git remote add origin git@github.com:ItalyPaleAle/svelte-journal.git
git push -u origin master
```

If this is your first time pushing to GitHub, you'll have to authenticate yourself. There are two ways to do that, as follows:

- **Using SSH**: You will first need to set up a SSH key, as per this documentation: `https://bit.ly/github-auth-ssh`.

- **Using HTTPS and a Personal Access Token** (PAT): Instructions to set this up are here: `https://bit.ly/github-auth-https`. (Note that you won't be able to use your GitHub account password; you need to use a PAT instead.)

After this is done, your code will be pushed to the GitHub repository, as illustrated in the following screenshot:

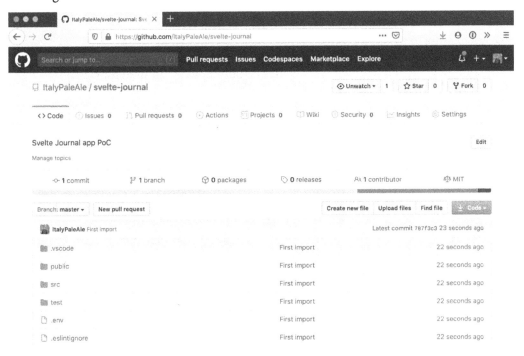

Fig 6.8 – Code pushed to GitHub

Creating a GitHub action to deploy the app

Lastly, we need to enable CI/CD by creating an action that automatically builds and deploys the app to Azure Storage on every commit to the `master` branch.

As mentioned, Actions is a CI/CD (and repository automation) platform that's built into GitHub itself, and it's free for public repositories (such as open source projects). Actions are defined using YAML files inside the `.github/workflows` folder (note the dot at the beginning of the folder's name).

Getting a service principal to authenticate with Azure

Our CI/CD pipeline will connect to our Azure environment to deploy the application in the Azure Storage account and to automatically purge the CDN. For the pipeline to do that, it needs to be authenticated.

In the Azure world, automation tools authenticate with an Azure subscription using a *service principal*. You can generate one with this command (no need to change anything in it):

```
az ad sp create-for-rbac \
   --name "Svelte-GitHub-Action" \
   --role Contributor \
   --sdk-auth
```

This will output a JSON document similar to this (output was truncated for clarity):

```
{
    "clientId": "…",
    "clientSecret": "…",
    "subscriptionId": "…",
    "tenantId": "…",
    "activeDirectoryEndpointUrl": "https://login.microsoftonline.
      com",
    "resourceManagerEndpointUrl": "https://management.azure.
      com/",
    …
}
```

Copy the entire JSON fragment, which we'll use in a moment. Note that it contains tokens that can be used to create and manage every single Azure resource in your subscription, so treat this like a password!

Setting up the secret in GitHub

After copying the preceding JSON fragment, go to the **Settings** page of your GitHub repository and select the **Secrets** tab.

Create a new secret named AZURE_CREDENTIALS and paste the JSON document as the value, then save the changes, as illustrated in the following screenshot:

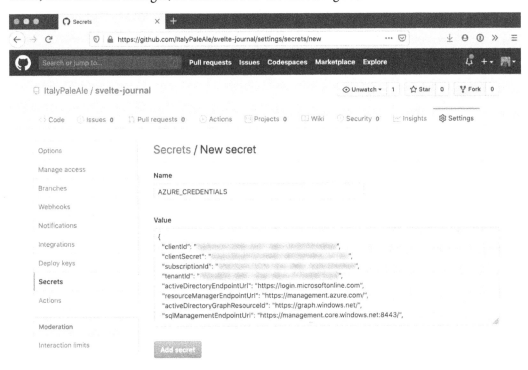

Fig 6.9 – Adding the AZURE_CREDENTIALS secret

Creating the action YAML workflow

Lastly, we are ready to build our Action. This will build, test (running the tests defined in *Chapter 5, Single-Page Applications with Svelte*), and deploy our application. Additionally, it will automatically trigger a purge of the CDN cache.

We need to create a file called `.github/workflows/build-and-deploy.yaml`. This is a fairly long document, so copy and paste it from this book's GitHub repository (from the `ch6` folder) into your own project, at the same path.

There are a few values that need to be set at the beginning of this file:

.github/workflows/build-and-deploy.yaml (fragment)

```
env:
  RESOURCE_GROUP: 'Svelte-App'
  STORAGE_ACCOUNT_NAME: 'sveltejournal'
  CDN_PROFILE_NAME: 'Svelte-App-CDN'
  CDN_ENDPOINT: 'sveltejournalcdn'
```

At the beginning of the file, we are defining a few environmental variables for our action, which you'll need to update based on the names of your Azure resources. The preceding values reflect the ones used in this chapter's samples. The environmental variables are listed as follows:

- `RESOURCE_GROUP` is the name of the resource group that contains our Azure CDN.

- `STORAGE_ACCOUNT_NAME` is the name of the Azure Storage account containing our app.

- `CDN_PROFILE_NAME` is the name of the Azure CDN profile.

- `CDN_ENDPOINT` is the name of the Azure CDN endpoint.

The rest of the file contains the various tasks that the pipeline needs to run, which you can read from in the document directly.

> **Working with GitHub Actions**
> The full documentation on creating GitHub actions is available here:
> `https://help.github.com/en/actions`.

After you have created the file, commit it (with `git add .github/* && git commit -m"Adding Action"` or by using the **SOURCE CONTROL** tools in Visual Studio Code), then push the changes to GitHub (`git push` or the equivalent command in Visual Studio Code).

As soon as the file is added to GitHub, the action will be triggered automatically, and it will run again on every commit on the `master` branch, re-building, re-testing, and re-deploying your application, automatically!

> **Tests failing because of chromedriver**
>
> If your action fails because tests fail, first check to make sure that the version of the `chromedriver` dependency in the `package.json` file is updated. Refer to the test setup in *Chapter 5, Single-Page Applications with Svelte,* for details.
>
> If you don't plan to use tests, you can remove the `test` script from the `package.json` file.

Making a code change

Now that we've set up CI/CD, we can test that by making a code change to get the app automatically updated.

For example, we could try changing the title shown in the navbar. In the `src/components/Navbar.svelte` file, change this line:

src/components/Navbar.svelte (fragment: before)

```
<i class="fa fa-pencil-square-o" aria-hidden="true"></i>
  Journal
```

The preceding line should be changed to this:

src/components/Navbar.svelte (fragment: after)

```
<span class="text-blue-800"><i class="fa fa-smile-o"
  aria-hidden="true"></i> My happy journal</span>
```

Commit the code change and push it to GitHub.

In the **Actions** tab for your GitHub repository, you will see the pipeline being triggered immediately and you can follow its progress, as in the next screenshot:

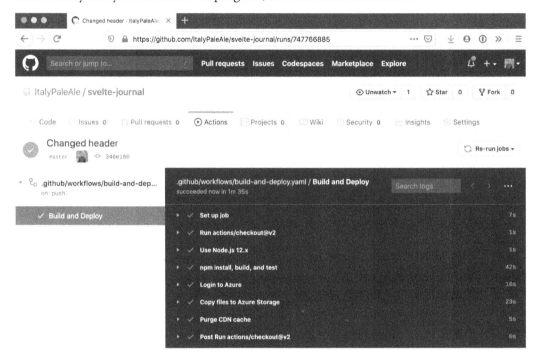

Fig 6.10 – The action ran successfully

After the pipeline is completed successfully, wait a few minutes for the CDN to finish purging (can take up to 10 minutes). Then, refresh your app to see the new header, as shown in the following screenshot:

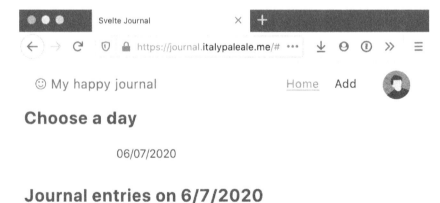

Fig 6.11 – The app's header was updated by the pipeline

Managing and destroying your resources

As a very last step, we should look at how to manage (and destroy) all the Azure resources we've created.

You can access the Azure portal at `https://portal.azure.com`, then sign in with your account.

From the sidebar on the left, choose **Resource groups** to list all of the resources, including the one we've created earlier, as shown below:

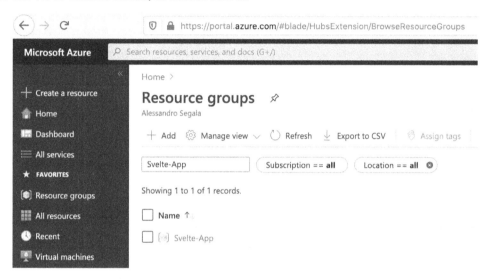

Fig 6.12 – The Resource groups view in the Azure portal

Clicking on the `Svelte-App` resource group (or whatever you've named it) will show all the resources, like in this screenshot:

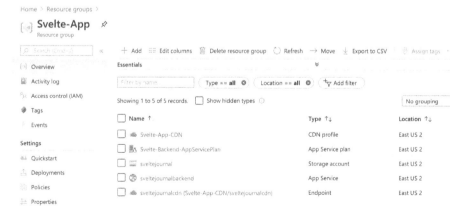

Fig 6.13 – View resources in the resource group

From there, you can manage all the resources we've created, including the storage account, the CDN, and the web app.

Additionally, you can shut down all your resources by pressing on the **Delete resource group** button at the top of the list. This will ensure that you won't be billed for unused resources anymore.

Summary

In this chapter, we've taken our application and brought it to production, deploying it to the cloud so that others can access it.

After exploring a variety of options to deploy JAMstack apps, we've learned how to use Azure Storage (an object storage service part of Microsoft Azure) for hosting and serving our app, and we've put a CDN in front of it to make our app load faster for all users worldwide.

We've also gone one step further and enabled CI/CD for our Svelte application, using GitHub and GitHub Actions. Now, our app is automatically built, tested, and deployed every time we make a code change.

This chapter concludes our work on building a PoC application with Svelte 3. In the next short chapter, we'll look at a list of resources for you to continue learning.

7
Looking Forward

At the end of *Chapter 6*, *Going to Production*, we completed our **Proof-of-Concept** (**PoC**) application with Svelte 3, deployed it to production with the Microsoft Azure cloud platform, and even implemented some DevOps practices with **Continuous Integration and Continuous Delivery** (**CI/CD**).

Looking back, you've done significant work, and we've come a long way together since the beginning of this book. Thank you for following along thus far! I hope you've learned a lot throughout this journey.

Of course, there are always new opportunities to continue learning and growing. To help you on your individual learning journey, this chapter contains some additional pointers, as follows:

- Where to find resources to further your understanding of Svelte and its ecosystem and community, including documentation, connections with other developers, third-party shared components, and support

- An overview of Sapper

- Some additional resources that you can leverage to build powerful apps with the JAMstack

The Svelte ecosystem and community

We mentioned at the beginning of the book that Svelte is one of the newer frameworks for frontend development with JavaScript: Svelte 3 (a complete rewrite from the previous versions) was released as recently as April 2019.

Because of that, as I'm writing this, the amount of resources for Svelte 3 is still smaller than what developers using "older" frameworks such as React or Angular can find: tutorials, articles, videos, third-party components, events, and even books.

However, the good news is that the Svelte community is growing fast, and an increasing number of passionate developers and users are creating more content every day.

Official websites and documentation

The official website for Svelte is `https://svelte.dev`.

In there, you can also find the complete documentation for the framework (`https://svelte.dev/docs`), as well as more interactive examples.

As mentioned in an earlier chapter, on the official website you can also play with the Svelte REPL, which is an interactive experience that lets you write Svelte components in the browser and see them getting compiled and run side by side: `https://svelte.dev/repl/`. The REPL is a really convenient tool for experimenting with Svelte and testing ideas. It is particularly useful to share code snippets with others too—for example, in case you run into issues with Svelte and need support (see the next sections).

Lastly, worth highlighting here is the official Svelte project on GitHub: `https://github.com/sveltejs/svelte`. This is where the development of the Svelte framework happens. If you encounter a bug in Svelte itself (rather than in your code), you can create an issue in there to flag it to the developers.

Svelte community website

More recently, the Svelte project created a website for its community: `https://svelte-community.netlify.app/`.

In there, you can find things like this:

- A list of reusable third-party components that you can add to your project. Both `svelte-calendar` and `svelte-spa-router`, the two external Svelte components we used in our PoC app, are listed there, alongside many others in a growing list that includes **User Interface** (**UI**) frameworks too.

- A list of events and meetups for Svelte developers.

- A list of miscellaneous resources, including blogs and recordings of conference talks about Svelte, such as those delivered by Rich Harris (the original creator of Svelte).

Community support and connections

If you have issues with your own applications that you're building with Svelte, you can find a welcoming group of developers in the community that can help.

Additionally, these can be good places to hang out if you want to chat with other developers working with Svelte and share knowledge and experiences.

You can connect with Svelte developers and get community support in these places online:

- **Discord**: The official Discord server is very active, and it's a good place to get real-time support from the community or just talk about Svelte. You can find the Svelte Discord server here: `https://svelte.dev/chat`.

- **Reddit**: On `/r/sveltejs`, you can find a variety of posts about Svelte, including help requests, links to resources (articles, videos, blogs, and components), and more: `https://reddit.com/r/sveltejs`.

- **Stack Overflow**: You can ask questions using the `svelte` and/or `svelte-3` tags to get help: `https://stackoverflow.com/`.

The Svelte project has also an official Twitter account, `@SvelteJS` (`https://twitter.com/sveltejs`), which you can follow to get updates on Svelte and learn about events, new resources, and so on.

For those who prefer podcasts, Svelte Radio (`https://www.svelteradio.com/`) is published on platforms including Apple Podcasts, Spotify, and Google Podcasts. It features interviews with developers using Svelte and community members, as well as maintainers of Svelte itself.

Lastly, if you're looking for a job as a developer working with Svelte, you can look at the official job board: `https://sveltejobs.dev/`.

Sapper

Sapper (`https://sapper.svelte.dev/`) is a full-featured application framework for building multi-page **Progressive Web Apps** (**PWAs**). Sapper is based on Svelte 3, which it uses to allow developers to build pages and components, and it is maintained as a sub-project of Svelte itself. Among developers working with Svelte, Sapper has a modest but growing fanbase. (We saw a brief mention of multi-page applications and PWAs in *Chapter 1, Meet Svelte*, and we compared them with **Single-Page Applications** (**SPAs**), such as the one we were about to start building, throughout this book.)

With Sapper, developers can write components using Svelte, just like we've seen in *Chapter 3, Building Reactive Svelte Components*, and *Chapter 4, Putting your App Together*. Sapper then automatically builds multi-page PWAs, enabling features such as **Server-Side Rendering** (**SSR**), routing (using a SEO-friendly router based on the HTML5 History API, which means it's better indexed by search engines), and offline support; Sapper is optimized for performance and for ease of development. On the downside, Sapper applications usually require some degree of server-side processing and can't be hosted on an object storage service like we've done in the previous chapter.

As you've become familiar with Svelte throughout this book, if the promises of Sapper sound interesting to you, I encourage you to check it out.

More JAMstack resources

When building apps using the JAMstack, you can leverage a variety of external APIs and **Software-as-a-Service (SaaS)** solutions that integrate with your applications running within the web browser. These APIs provide identity, access to data, storage, and other services.

In this section, we'll look at some tools and services you can leverage, but this list is far from complete (and apologies in advance if I did not include your favorite tool or service). Note in the following list that each section is ordered alphabetically:

- **Static site generators**

    These are normally used as part of your CI/CD pipeline, and include the following:

    Gatsby: `https://www.gatsbyjs.com/`

    Hugo: `https://gohugo.io/`

    Jekyll: `https://jekyllrb.com/`

- **Authentication services (supporting OpenID Connect)**

  Amazon Cognito: `https://aws.amazon.com/cognito/`

  Auth0: `https://auth0.com/`

  Azure AD: `https://azure.microsoft.com/services/active-directory/`

  Azure AD B2C: `https://azure.microsoft.com/services/active-directory/external-identities/b2c/`

  Firebase Authentication: `https://firebase.google.com/products/auth`

  Google Identity Platform: `https://developers.google.com/identity`

  Keycloak (self-hosted): `https://www.keycloak.org/`

  Okta: `https://www.okta.com/`

  ORY (self-hosted): `https://www.ory.sh/`

- **Social authentication (supporting OpenID Connect)**

  Apple (Sign in with Apple): `https://developer.apple.com/sign-in-with-apple/`

  Facebook: `https://developers.facebook.com/docs/facebook-login/`

  Google Identity Platform: `https://developers.google.com/identity`

  Microsoft identity platform: `https://docs.microsoft.com/azure/active-directory/develop/v2-protocols-oidc`

- **Content Management Systems (CMS)**
  All these options can be self-hosted, but some are offered as cloud services too. They include the following:

  Ghost: `https://ghost.org/`

  Netlify CMS: `https://www.netlifycms.org/`

  Strapi: `https://strapi.io/`

  WordPress (REST APIs):
  `https://developer.wordpress.org/rest-api/`

- **Databases**

  Cloud Firestore: `https://firebase.google.com/products/firestore/`

  CouchDB (self-hosted): `https://couchdb.apache.org/`

- **E-commerce and payments**

  Flatmarket (self-hosted): `https://github.com/christophercliff/flatmarket`

  PayPal: `https://developer.paypal.com/`

  Shopify Headless commerce: `https://www.shopify.com/plus/solutions/headless-commerce`

  Stripe: `https://stripe.com/`

Additionally, for code that must be executed in a server, you can leverage *serverless* platforms, which provide an environment to run applications dynamically and can be invoked by JAMstack apps, or even *no-code* services, which allow integrations to be built without writing any code at all.

Examples of **serverless** platforms include the following:

- AWS Lambda: `https://aws.amazon.com/lambda/`
- Azure Functions: `https://azure.microsoft.com/services/functions/`
- Cloudflare Workers: `https://workers.cloudflare.com/`
- Google Cloud Functions: `https://cloud.google.com/functions`

Among **no-code** services, here are the ones worth mentioning:

- Azure Logic Apps: `https://azure.microsoft.com/services/logic-apps/`
- Microsoft Power Automate: `https://flow.microsoft.com/`
- Zapier: `https://zapier.com/`

With regard to hosting our JAMstack app itself, we've explored a variety of options in *Chapter 6, Going to Production.*

Summary

In this last chapter, we have looked at a list of resources that we can leverage to continue learning about Svelte, connecting with the community, and obtaining help with our code.

We have also looked at a (fairly long, yet absolutely incomplete) list of tools and services that we can use to build powerful JAMstack apps that run within a web browser.

This completes our journey together. I hope you found this book helpful to learn about Svelte and, after building this PoC app together, I hope you will be able to create your own static web apps with Svelte 3.

Thank you for reading!

Other Books You May Enjoy

If you enjoyed this book, you may be interested in these other books by Packt:

Hands-On JavaScript High Performance

Justin Scherer

ISBN: 978-1-83882-109-8

- Explore Vanilla JavaScript for optimizing the DOM, classes, and modules, and querying with jQuery
- Understand immutable and mutable code and develop faster web apps
- Delve into Svelte.js and use it to build a complete real-time Todo app
- Build apps to work offline by caching calls using service workers
- Write C++ native code and call the WebAssembly module with JavaScript to run it on a browser
- Implement CircleCI for continuous integration in deploying your web

Clean Code in JavaScript

James Padolsey

ISBN: 978-1-78995-764-8

- Understand the true purpose of code and the problems it solves for your end-users and colleagues

- Discover the tenets and enemies of clean code considering the effects of cultural and syntactic conventions

- Use modern JavaScript syntax and design patterns to craft intuitive abstractions

- Maintain code quality within your team via wise adoption of tooling and advocating best practices

- Learn the modern ecosystem of JavaScript and its challenges like DOM reconciliation and state management

- Express the behavior of your code both within tests and via various forms of documentation

Leave a review - let other readers know what you think

Please share your thoughts on this book with others by leaving a review on the site that you bought it from. If you purchased the book from Amazon, please leave us an honest review on this book's Amazon page. This is vital so that other potential readers can see and use your unbiased opinion to make purchasing decisions, we can understand what our customers think about our products, and our authors can see your feedback on the title that they have worked with Packt to create. It will only take a few minutes of your time, but is valuable to other potential customers, our authors, and Packt. Thank you!

Index

S

www.ingramcontent.com/pod-product-compliance
Lightning Source LLC
Chambersburg PA
CBHW060505090326
40690CB00069B/5108